Anthropological Papers
Museum of Anthropology, University of Michigan
No. 88

Elements for an Anthropology of Technology

by
Pierre Lemonnier

with a foreword by John D. Speth

Ann Arbor, Michigan
1992

© 1992 by the Regents of The University of Michigan
The Museum of Anthropology
All rights reserved

ISBN 978-0-915703-30-2 (paper)

ISBN 978-1-949098-84-6 (ebook)

Cover design by Katherine Clahassey

The University of Michigan Museum of Anthropology currently publishes three mono-graph series: Anthropological Papers, Memoirs, and Technical Reports. We have over seventy titles in print. For a complete catalog, write to Museum of Anthropology Publications, 4009 Museums Bldg., Ann Arbor, MI 48109-1079, or call (313) 764-0485.

Library of Congress Cataloging-in-Publication Data

Lemonnier, Pierre, 1948–
   Elements for an anthropology of technology / Pierre Lemonnier :
with a foreword by John D. Speth.
     p. cm. — (Anthropological papers / Museum of Anthropology,
University of Michigan ; no. 88)
   Includes bibliographical references.
   ISBN 0-915703-30-0 (acid-free : pbk.)
   1. Technology and civilization. 2. Philosophical anthropology.
I. Title. II. Series: Anthropological papers (University of
Michigan. Museum of Anthropology) ; no. 88.
GN2.M5 no. 88
[T14]
306 s—dc20
[303.48'3]   92-19485
                                                              CIP

Contents

List of tables	v
List of figures	v
Foreword, by John D. Speth	vii
Preface	xi

CHAPTER 1. TECHNOLOGY AND ANTHROPOLOGY — 1
    Introduction — 1
    The What and Why of an Anthropology of
        Technological Systems — 4
    Classical Anthropological Approaches to Technology — 11
    Economic Anthropology and Technological Systems:
        A One-Sided Approach — 14
    A Complementary Approach — 17

CHAPTER 2. FROM FIELD TO FILES: DESCRIPTION AND
ANALYSIS OF TECHNICAL PHENOMENA — 25
    Operational Sequences: The Basic Data — 25
    What and How to Classify — 32

CHAPTER 3. ARBITRARINESS IN TECHNOLOGIES — 51
    Social Representation of Technologies Among
        the Anga — 51
    A Glance at Airplanes: Arbitrary Choices in So-Called
        High-Tech Society — 66

CHAPTER 4. SOCIAL REPRESENTATION OF TECHNOLOGIES — 79
    Technological Traits as Signs — 79
    Style and Function — 85
    Style Without Function? — 89
    Technology and Information — 91
    Excesses of the Cambridge School — 96

CHAPTER 5. CONCLUSIONS: TOWARD A STUDY OF SOCIAL
REPRESENTATIONS OF TECHNOLOGY   105
    Back to the Anga Case   105

Notes   117

References Cited   119

List of Tables

1. Comments on a tree diagram, 32
2. Frequency and ways of stirring brine in salt marshes in France, 44
3. Technological distribution of principal types of agriculture throughout the world, 45
4. Classification of Dani arrows, 46
5. Types of percussion, 47
6. Functional hierarchy of Moravian folk costumes, 94
7. Summary of an operational sequence: two men at work, followed by one woman, 108

List of Figures

1. Operational sequence ("work chain"): The basic model, 31
2. A tree diagram in which each set of branches represents the details of the operation immediately preceding, 31
3. Two simplified operational sequences, 33
4. Description and analysis of an operational sequence, 34
5. Diagram of an operational sequence, 35
6. Spinning technologies in Gîlan and Azerbaijan, 39
7. Types of houses in Tuscany, 40
8. Types of houses of North American Indians, 41
9. Distribution of conical and subconical houses of North American Indians, 42
10. Dominant house-building division of labor among North American Indians, 43
11. Classification of planes according to number of struts, control surface, and center of gravity, 48
12. Steps in salt production in New Guinea, 49
13. Arrows with grooves to allow breakage, 53
14. Nature and variability of Anga technical traits, 54
15. Barbed arrows, 55
16. Types of Anga hearths placed directly on floor, 56
17. Pit trap for pigs, 57
18. Pit trap constructed at opening in fence, 58
19. Dead-fall trap for pigs, 59
20. Distribution of barbed arrows, 60
21. House with single enclosure wall, 60
22. House with double enclosure wall, 62
23. Distribution of houses with single and double enclosure walls, 64
24. Distribution of bow types, 65
25. Engine between wings of multiplane, 68
26. Engine at back of fuselage, 68

27. Engines inboard on wings and included in fin, *68*
28. Passengers in floats of hydroplane, *69*
29. Passengers in two fuselages, *69*
30. Passengers aboard a "flying wing," *70*
31. Canard foreplane, *71*
32. Phillips multiplane, *72*
33. Dixon's Nipper, *72*
34. Cessna C137 Skymaster, *74*
35. Mitsubishi MU-2, *74*
36. Mach 2 planes, *76*
37. World War II fighter and 1930s racing hydroplane, *76*
38. Bark cape and skirt, *107*
39. Stone beater for pounding bark, *109*

Foreword

John D. Speth

The study of technology has a long and somewhat checkered history in anthropology. In the nineteenth and early twentieth century, such studies were very fashionable, being eagerly undertaken by anthropologists as they sought to document the seemingly endless variety of "primitive" human cultures that were being discovered in the remotest corners of the world. Unfortunately, most of these early studies provided what in hindsight seem rather sterile and largely descriptive inventories of material culture rather than explorations of how technologies in these societies were organized and how they were integrated into the larger social, economic, and symbolic whole. It is not surprising, therefore, that as the discipline of anthropology matured, interest rapidly shifted away from narrow studies of material culture to new domains, such as kinship, sociopolitical organization, religion, and ideology, research topics that promised to offer much broader, more comprehensive, and more relevant views of human culture.

The one subfield within anthropology that has maintained a lively interest in material culture and technology is of course archaeology. While archaeologists have generally sought to deal with the same phenomena that attract the energies of their colleagues in cultural anthropology, getting at past kinship systems, political organizations, or religious beliefs is far more difficult when there are no living informants. Instead, archaeologists have had to devise ingenious ways to get at these phenomena indirectly through the spatial and temporal patterning of material culture remains. The consequence, though, has been that archaeologists often look at material culture not as a way of studying past technologies in their own right, but as an indirect way to accomplish other objectives, such as ordering sites in time, or reconstructing past subsistence practices, identifying status differences, or establishing "cultural" affiliation.

Even archaeologists who concern themselves with issues of style, such as the variability in form or design on an object, are often using style as a proxy to assist in identifying other behavioral phenomena in the past, such as patterns of inter-group interaction, competition, or boundary maintenance. They make, as Lemonnier points out, no explicit attempt "to explain why one given aspect of material culture (e.g., the shape of pots), and not another (e.g., the way people walk, dress, or hunt) is used to express certain social relations. . . ." In other words, while archaeologists are very interested in many facets of technology, and now often undertake their own ethnographic fieldwork in order to gain insights into specific classes of material culture, like most of their colleagues in ethnology they devote very little systematic effort toward developing an anthropology of technology.

The present volume by Pierre Lemonnier, a renowned French anthropologist, takes a refreshing new look at technology, one that will be of great interest to ethnologists and archaeologists alike. Lemonnier draws on examples from tribal societies in New Guinea, where he has done extensive fieldwork, and from the contemporary aeronautics industry in our own society, to show that technologies, even highly sophisiticated ones, involve many arbitrary decisions and choices that are not dictated by immediate material or physical constraints, but instead reflect higher-level systems of meaning within the society. By showing that technologies, be they "primitive" or modern, participate in complex systems of meaning that are intimately bound up in the larger symbolic system, Lemonnier elegantly pulls the study of technology back into the mainstream of contemporary ethnology.

Lemonnier also considers issues that are of considerable interest to archaeologists. For example, he offers an insightful and timely critique of the widespread view among contemporary prehistorians, growing out of the work of Wobst, Wiessner, and others, that stylistic variability serves largely as a means of conveying information about social identity. He also critiques symbolic and structural archaeologists, such as Hodder and Tilley, who see style primarily in terms of power relations. He rightly points out that these perspectives, while useful and interesting, are overly narrow, because they usually single out only one or at best only a few items of material culture (e.g., pots, headbands, or arrowheads), and then attempt to explain the observed variability in isolation from other components of the technology with which the selected items are undoubtedly linked. Moreover, the meaning of any particular style is ultimately tied to higher-level systems of meaning within a society and therefore cannot be understood or explained in isolation. Finally—even if one does demonstrate that style in a given context, for example the decoration of a pot, does reflect social identity or social stratification—current archaeological studies still lack a body of theory that attempts to predict why that particular medium (i.e., the pot) was selected to convey this kind of information in one society but not in another.

Lemonnier's study does not pretend to offer definitive answers to all of the many interesting questions it raises. But ethnologists and archaeologists will find here an extremely useful framework and methodology for furthering the development of an anthropology of technology. The reader will also find that French anthropologists have already made considerable headway in this important endeavor. However, because of unfortunate language barriers much of this interesting work has remained unknown to most North American anthropologists. It is truly a pleasure, therefore, to be able to publish this stimulating, new, and challenging work by one of France's leading specialists on the anthropology of technology. Our hope is that the appearance of this volume will spark a renewed interest in technology among English-speaking ethnologists and archaeologists, one that helps to reintegrate the holistic study of technology back into mainstream anthropology.

Preface

This small book is neither a manual nor an ethnographic monograph. Instead, it summarizes some general theoretical and methodological issues with which I have been dealing ever since I began, more than fifteen years ago, to study traditional salt-making in France and the Anga people of Papua New Guinea, and became intensely interested in the ethnology of technological systems. Linking technology and society from an anthropological point of view requires that we delimit the domain of human material life, define the kinds of questions we wish to address to material culture, and propose suitable theoretical and methodological means for answering these questions. The answers I propose here are far from definitive; instead, they reflect the current state of research which is still ongoing, and should serve primarily as guides to help us progress further.

Although the answers I present here are deeply embedded in examples from my own research, in reality they are part of a collective research tradition which started in France more than fifty years ago. Marcel Mauss, André Leroi-Gourhan, André-Georges Haudricourt, and Bertrand Gille were pioneers whose thoughts and writings still heavily influence present-day research dealing with technology. While I was influenced by the first two scholars only through their writings, I have been fortunate to have had many conversations with Professor Haudricourt, where we discussed topics ranging from the existence of technological "choices" to various aspects of Melanesian pig husbandry. I also had the exciting experience of reading Bertrand Gille's analysis of Greek, Roman, and Medieval technology while I was living among New Guinean horticulturalists, which gave me the sensation (obviously an incorrect one) that what I was witnessing in Melanesia was somehow a fleeting glimpse of an earlier time. I was also fortunate to have met Professor Gille personally a few months before he died, and although our visit was only for two hours, I am still touched, years later, by the affability, humility, and generosity of this scholar who combined a powerful theoretical perspective with truly encyclopedic knowledge.

My work has also been deeply influenced by Robert Cresswell and his CNRS research team in Paris, "Techniques et Culture," to which I have belonged since 1972. Needless to say, much of what I am writing here has been discussed in our internal seminars, and I will often express ideas here that are as much theirs as mine. François Sigaut, a companion and good friend for many years, has also profoundly influenced my thinking, and although we disagree as much as we agree, our divergent views have provided important and stimulating challenges.

The following chapters took shape during a series of lectures called "The Ethnology of Technology" which I had the good fortune to present as a "Chargé de Conférence" in the Ecole des Hautes Etudes en Sciences Sociales (Paris) in 1983–1986. The volume took final shape in the fall of 1986 at the Museum of Anthropology of the University of Michigan, where I was invited at the initiative of Robert Whallon to spend a semester in residence. Indeed, I found ideal conditions there to work on this project, and benefitted from the kind and lively comments of the graduate students who attended a weekly seminar of mine. I was also fortunate to have Karl Hutterer as a participant in the seminar, and John Speth who regularly checked my "English" a few hours before my weekly presentations, and did not know yet that he would spend many hours and days editing the present book. Both of them had much more important things to do than taking care of me. I am very thankful to the Department and Museum of Anthropology, and to my friends there: I owe them all the happy days I spent there and the work that I was able to accomplish.

P.L.
Paris, 1989

Chapter 1

Technology and Anthropology

INTRODUCTION

The anthropology of technological systems, or the study of material culture in a social and economic context, is still a new discipline. The field could have developed more than fifty years ago, when Marcel Mauss (1935) showed that some of our most casual acts, in which our body alone is involved, such as standing up, sitting down, sleeping, walking, or swimming, are culturally determined. If such seemingly "natural" behaviors were in fact highly socialized, it then seemed obvious to him that more complex actions which involved tools or other objects were the product of social learning processes. Perhaps Mauss thought his demonstration was clear enough to launch anthropology into the study of technological behavior. But this has not been the case.

At this point, I'd like to propose a provisional definition of technology; a more detailed one will be given later. Technology embraces all aspects of the process of action upon matter, whether it is scratching one's nose, planting sweet potatoes, or making jumbo jets.

Technologies are not only things and means used by societies to act upon their physical environment. For the ethnologist, and for the archaeologist and historian as well, technologies are—like myths, marriage prohibitions, or exchange systems—social productions in themselves. As Conklin (1982:16) put it, technologies are the "material expression of cultural activity." And yet the social dimension of technological action—that is, why and how a given society uses a particular technology and not another—is rarely taken into account by anthropologists. Anthropologists rarely ask questions such as: what is the social context of a technological "choice"?; or in what respect is a technology, any technology, a social production? These might be the first questions anthropologists should ask about action on matter. Other important questions would be: why, if all other things are equal, do societies adopt certain technological features and reject others?; to what extent do these technological choices influence transformations of technological systems and societies?; and how are these choices compatible with other social choices? Clearly, an anthropology of technology must, besides providing inventories of technologies and serving as a complement to the study of the *effects* of technologies on societies, also deal with the relationship between technological systems and other social phenomena.

Of course, anthropologists and archaeologists have been studying material culture, or technological systems, for years. Museums are full of artifacts, with or without the right tags. Any good traditional monograph contains a sketch of common house types and, quite often, the form and capacity of traditional teapots, plow shapes and lengths, and so forth. We even have a few descriptions of what goes on when planting a garden, or when building a house, or on a wild boar hunt, complete with time charts, maps, and pictures. Some fundamental results in economic anthropology have been produced by minute comparative studies of the productivity of stone versus steel tools. Clearly, then, anthropology has a long-standing interest in material culture, and some scholars have shown an awareness of key links between particular aspects of material culture and important social features. More generally, we all know that historians and economists have shown at least that the potter's wheel, the water mill, and the steam engine have something to do with the productivity of work. Archaeologists have devoted thousands of pages to describing and analyzing the decorations on many types of artifacts, and some researchers have tried to relate these analyses to other aspects of the society that produced or used the artifacts. Often the latter approach appeals to a theory of information. The results are quite good when informants are asked directly to say something about the color of their neckties or the shape of their teapots, but less satisfactory when it is the designs or decorations themselves which are asked to say something, especially something relevant to current

Chapter 1

Technology and Anthropology

INTRODUCTION

The anthropology of technological systems, or the study of material culture in a social and economic context, is still a new discipline. The field could have developed more than fifty years ago, when Marcel Mauss (1935) showed that some of our most casual acts, in which our body alone is involved, such as standing up, sitting down, sleeping, walking, or swimming, are culturally determined. If such seemingly "natural" behaviors were in fact highly socialized, it then seemed obvious to him that more complex actions which involved tools or other objects were the product of social learning processes. Perhaps Mauss thought his demonstration was clear enough to launch anthropology into the study of technological behavior. But this has not been the case.

At this point, I'd like to propose a provisional definition of technology; a more detailed one will be given later. Technology embraces all aspects of the process of action upon matter, whether it is scratching one's nose, planting sweet potatoes, or making jumbo jets.

Technologies are not only things and means used by societies to act upon their physical environment. For the ethnologist, and for the archaeologist and historian as well, technologies are—like myths, marriage prohibitions, or exchange systems—social productions in themselves. As Conklin (1982:16) put it, technologies are the "material expression of cultural activity." And yet the social dimension of technological action—that is, why and how a given society uses a particular technology and not another—is rarely taken into account by anthropologists. Anthropologists rarely ask questions such as: what is the social context of a technological "choice"?; or in what respect is a technology, any technology, a social production? These might be the first questions anthropologists should ask about action on matter. Other important questions would be: why, if all other things are equal, do societies adopt certain technological features and reject others?; to what extent do these technological choices influence transformations of technological systems and societies?; and how are these choices compatible with other social choices? Clearly, an anthropology of technology must, besides providing inventories of technologies and serving as a complement to the study of the *effects* of technologies on societies, also deal with the relationship between technological systems and other social phenomena.

Of course, anthropologists and archaeologists have been studying material culture, or technological systems, for years. Museums are full of artifacts, with or without the right tags. Any good traditional monograph contains a sketch of common house types and, quite often, the form and capacity of traditional teapots, plow shapes and lengths, and so forth. We even have a few descriptions of what goes on when planting a garden, or when building a house, or on a wild boar hunt, complete with time charts, maps, and pictures. Some fundamental results in economic anthropology have been produced by minute comparative studies of the productivity of stone versus steel tools. Clearly, then, anthropology has a long-standing interest in material culture, and some scholars have shown an awareness of key links between particular aspects of material culture and important social features. More generally, we all know that historians and economists have shown at least that the potter's wheel, the water mill, and the steam engine have something to do with the productivity of work. Archaeologists have devoted thousands of pages to describing and analyzing the decorations on many types of artifacts, and some researchers have tried to relate these analyses to other aspects of the society that produced or used the artifacts. Often the latter approach appeals to a theory of information. The results are quite good when informants are asked directly to say something about the color of their neckties or the shape of their teapots, but less satisfactory when it is the designs or decorations themselves which are asked to say something, especially something relevant to current anthropological questions concerning, for example, political power, male–female relationship, and so on.

Despite the concern of anthropologists and archaeologists with portions of technological systems, the conclusions they come to are generally disappointing. Even with the right tags, artifacts alone do not have much to say, as we shall see later. And the few economic anthropologists who are concerned with material culture usually are satisfied with knowing the potential output or the immediately apparent form of the organization of work.

In most cases, technological systems are summed up merely as static *constraints* without considering the social aspects of material culture. And in the few cases where the social aspects are explored, technological systems are reduced to statements about the shape of artifacts or, worse, their decoration: in other words, to their informational dimension. Action on matter is nearly always left aside.

Yet a social theory of material culture should deal with technologies in their most physical aspects, that is to say, with the way they are made and used for some action on the material world. In point of fact, the color of a digging stick, the shape of a grass skirt, or the decoration on a throwing spear, are not of much importance in harvesting taro, protecting one's body from cold, or hunting kangaroos (at least as long as the possible magical powers of the decoration are not included). And yet ninety-nine percent of the studies that do take into account some social dimension of material culture ignore the physical action of technologies on the material world, as though Mauss had never written his "Les Techniques du Corps" ("On Body Techniques") in 1935.

We should not concentrate just on the immediate or obvious informational aspects of material culture (such as styles of costumes, decorative motifs, colors). There are more subtle informational or symbolic aspects of technological systems that involve arbitrary choices of techniques, physical actions, materials, and so forth that are not simply dictated by function, but which are integral components of the larger symbolic system. We shall discover that some of these technological features which are directly involved in actions on the material world are also objective indicators of meaning and, as such, can be called "symbols." These more subtle aspects must be found not just by "reading" style, but through synchronic or diachronic analyses of the relations among elements of a technological system, their transformations, and their social representations at levels beyond the mere realm of actions on the material world.

Step by step in the chapters that follow, I shall try to demonstrate what up to now I have merely been suggesting. I first define what technologies are made of, and outline their basic features, particularly how technologies are systems. Then I suggest some questions that may be central to an anthropology of material culture. For that purpose, I survey and comment upon the usual encounters between technology and anthropology, as well as history, and outline a complementary approach, one already alluded to above. Chapter 2 is devoted to methodological problems of definition, observation, description, transcription, and analysis, of "operational sequences," which are the basic data of an anthropology of technological systems. As there is no area of ethnology or archaeology that can function without comparison, we have to decide what to classify in our new data, and how to do it in order to make productive comparisons. For this, I turn to Leroi-Gourhan, better known as a prehistorian, whose methods and approaches are fundamental to an anthropology of technological systems. In Chapter 3, I consider the arbitrary nature of technological phenomena, using as case studies the material culture of the Anga tribes of Highland Papua New Guinea and the history of modern aviation.

From there, I attempt in Chapter 4 to offer some explanation of the various kinds of technological choices. The seemingly straightforward act of lacing one's shoes, when we examine the act in more detail, is conditioned and determined by a great many social factors. So is the choice a society makes to use traps rather than bows and arrows. (I call these social factors of technological action "social representation," a concept I will define more precisely later in Chapter 4).

Also in Chapter 4, I review some ethnological and archaeological theories, even though they are mainly concerned with stylistic, or informational, aspects of material culture, in order to test their usefulness for the study of technological systems. The results are not very satisfactory, and in Chapter 5 I suggest some directions of future research, using Anga data as an example, and urge that existing ethnographic or archaeological data on material culture be subjected to "new" questions (already asked by Leroi-Gourhan forty years ago): how are arbitrary "choices" in material culture to be identified, and where within the cultural system are they situated? In other words, style and function revisited.

THE WHAT AND WHY OF AN ANTHROPOLOGY OF TECHNOLOGICAL SYSTEMS

In his paper on body techniques, Mauss defined a technique as "an action which is *effective* and *traditional* (and in this it is no different from a magical, religious, or symbolic action) felt by the [actor to be] mechanical, physical or physico-chemical ... and ... pursued with this aim in view" (1935 [1979:104]). "Action" here refers to purposeful body movements and does not need any particular comment. "Traditional" means that these movements are inherited from the past and diversely "learned" by people. It follows that techniques are social phenomena, which may vary from one culture to another. "Effective" means that the material result obtained through technological action is the one that was sought. (Since everyday life shows that results sometimes differ from what was anticipated, maybe it is more accurate to say that "effective" merely means that the gesture seeks some physical result.) It has to be noted that Mauss's reference to the physical world does not mean that religious or magical thoughts or gestures are excluded from the technological domain. This raises the question of rituals, which are often aimed at, and linked to, effects in the physical world. I shall get around this difficulty by suggesting that to be called "technological," an action needs to involve at least some physical intervention which leads to a real transformation of matter, in terms of current scientific laws of the physical world.

I submit that every technique has five related components:

1) *Matter*—the material, including one's own body, on which a technique acts (e.g., clay, water, iron, sweet potatoes, aluminum).

2) *Energy*—the forces which move objects and transform matter.

3) *Objects*, which are often called artifacts, tools, or means of work. These are "things" one uses to act upon matter: a hammer, hook, steam-roller, or artificial salt-pond. It must be noted that "means of work" includes not only things that can be held in the hand; a factory is as much a means of work as is a chisel.

4) *Gestures*, which move the objects involved in a technological action. These gestures are organized in sequences which, for analytical purposes, may either be subdivided into "sub-operations" or aggregated into "operations" and then into "technological processes." I shall henceforth speak of "operational sequences," without any reference to a particular level of description.

5) *Specific knowledge*, which may be expressed or not by the actors, and which may be conscious or unconscious. This specific technological knowledge is made up of "know-how," or manual skills. The specific knowledge is the end result of all the perceived possibilities and the choices, made on an individual or a societal level, which have shaped that technological action. I call those possibilites and choices *social representations*. Some examples of social representations which shape a technology or technological action are: (a) the choice to use or not use certain available materials; (b) the choice to use or not use certain previously constructed means of action on matter (a bow and arrow, a car, a screwdriver); (c) the choice of technological processes (i.e., sets of actions and their effects on matter), and the results of these processes (e.g., a cooked meal, a house, or recently killed game); and (d) the choice of how the action itself is to be performed (a conception that it is the woman's role to cut firewood, or the man's to make fences for gardens).

I'd like to briefly comment on a few of these components listed above, though all will be addressed in greater detail later.

First, one must not forget the matter itself. By its own specificities and, of course, by being present or absent in a given environment, the materials may partially determine the technological behavior of a people. But a material may exist in the society's environment and yet remain unused. This means that we must study the knowledge people have of their own natural environment—in particular, the implicit or explicit classifications they apply to the materials available to them. Here we are approaching the core of classical anthropological research. It should be noted that reconstructing a people's classifications of materials is not beyond the reach of archaeologists, at least in some contexts.

Artifacts should be taken for what they are—only one part of technology. Yet the bulk of studies on technology has been devoted to artifacts, and only to artifacts. As most of these studies have been done by archaeologists, this is easy to understand. But interestingly, it was a prehistorian, Leroi-Gourhan (1943:43ff), who first drew our attention to the futility of looking at artifacts alone without considering the gestures that moved them.

There is a branch of anthropology that deals with gestures, but it is mainly concerned with immediately meaningful gestures such as those of the storyteller, the dancer, the person marching (Efron 1941; Calame-Griaule 1977; Cresswell 1968; Koechlin 1972; Polhemus 1978), or with the search for universals from the perspective of human ethology (Eibl-Eibesfeldt 1972). Until very recently, scholars have often shunned the study of gestures, simply because they were too hard to describe. However, new possibilities now exist for describing gestures, which link video recordings to computer analyses of simplified images. Thus, detailed study of technological gestures has become a reality, and its application is only a question of time (Abel 1984; Bril 1986; Bril and Sabatier 1986; Pelosse 1956, 1981).

Social representations of technologies are the channel through which social phenomena influence technological systems. Alongside the physical constraints presented by the material world available to a given society, social representations of technologies, too, are responsible for making and transforming technological systems.

Wobst (1977:32) made a fundamental remark, which is most important for our purpose, concerning the threefold nature of any technique: "material culture ... participates in and enhances exchanges of energy, matter and information in the human population that fashions it." We may pass over the distinction between matter and energy; however, the dividing line between physical aspects and informational aspects of material culture is crucial to our subject and difficult to explore. I shall come back to this point later, but it must be noted that my purpose here is not to identify where "function" stops and "style" begins (since style can be shown to have a function); rather, it is to investigate how, and to what extent, both physical functions and informational functions are interrelated in any technology. I maintain that informational functions may be found among the actual physical features of a technological system and not just in the so-called "stylistic" features which have little or no physical action on matter.

The next point is largely methodological in nature: it is very difficult to define or delimit a particular technique. For instance, gardening in a given society of New Guinea is a technique. Building a fence to protect a New Guinean garden from semi-domestic pigs is also a technique, a part of the first one. To sink a post in the ground, or to compress the ground at the bottom of the post with one's foot, heel, or toes, are also technological actions that might be called "techniques," and each can be isolated as a single technique. Selecting the appropriate level of description remains a problem and one that the researcher himself or herself must address on a case by case basis by deciding on a delimitation that fits his or her specific research problems.

The last feature of technologies to which I would like to draw attention is their systemic aspects. Long ago anthropologists started talking about technological systems in the same way as they talk about kinship systems or economic systems; that is, as arbitrarily delimited parts of a total social system. This was the position of Mauss (1947:29), for instance, and more recently of Lévi-Strauss (1976:11): " ... even the simplest techniques in any given society show systemic features." But the expression "technological system" is also quite common among engineers dealing with technology, for example, in the application of operational research to industrial problems during World War II. It is also frequently used by historians of technology, following the works of the late Bertrand Gille (1966, 1978, 1980).

Technological systems may be discussed at three different levels. First, we can discuss how the five components delineated above interact with each other to form a technology. Thus, gestures and knowledge are adapted to the physical evolution of the material being worked; a change in tools usually involves a change in technological knowledge and gestures; gestures are constantly being adapted to the dynamics of artifacts and changes in material, and so forth. If one of the components changes, in most cases the four others will have to change as well.

Second, if we now consider all of the technologies in a given society, it can easily be shown that most of them are interrelated. For one thing, a given technique often uses as raw materials the results of other techniques. Tools that are the result of the activities of steel factories are used by a carpenter to cut boards, which come together with other components (nails, glue, paint), which are the result of still other technologies. Technologies in one society may also be related because they share the same actors, the same places, the same artifacts, the same materials, the same sequences of gestures, or the same technological processes. The sharing by the actors in a given society of more or less the same social representations of technological behaviors is an important feature of the systemic aspect of a society's technologies.

It is the work of ethnologists and archaeologists to discover which techniques are related to which other techniques, and how (Lemonnier 1983). Studying houses in a New Guinea society, for example, necessitates examining and comparing different kinds of houses—women's houses, men's houses, ceremonial houses, pig shelters, garden houses, poultry coops, toilets built to meet patrol officers' requirements, stores, patrol officers' rest huts, and the ethnologist's house. But it is also necessary to study the relations between the walls or blinds in certain types of houses and garden fences, which may share the same materials or the same construction techniques (Steensberg 1980). At the same time, of course, fences cannot be studied without looking at gardening, and so on. In passing, this should make us aware of how little chance there is of understanding the material culture of any society by studying just a few artifacts, or, worse, by studying artifacts of only a single type. It should also be realized that the ethnography of technological systems is not a part-time occupation; it is as time-consuming to properly describe a technological system as it is to describe a kinship system or any other specialized area of anthropology.

The third level of discussion is the relation between technologies and other social phenomena. This level is already partially included in the previous one. Here we concern ourselves with how technological systems are integrated into the bigger systems we call societies. It is here that we define the general aims of an anthropology of technological systems.

Although frequently applied by economists and engineers to the study of particular questions in industrial societies, the systemic approach to technologies is rarely used for the study of so-called "primitive" or pre-industrial societies, and remains a way of viewing things rather than a developed methodology. Pioneering work on these matters was done by Bertrand Gille. Author of studies on Greek engineers (1980) and the technological revolution of the Renaissance (1966), and an expert on the technology of iron metallurgy since the Middle Ages (1970), Gille painstakingly developed the concept of technological systems in the more than one thousand pages of his *Histoire des Techniques* (1978), which he edited and for which he wrote fifteen of the nineteen chapters.

Gille used the concepts of "coherence" and "compatibility" in explaining the transformations of technological systems themselves, and of technological systems as related to other "systems" (which he termed "economic," "jural," "scientific"). For Gille, the history of technology was the history of the setting up and evolution of successive technological systems, from the Neolithic to the present. He thought that a given technique or set of techniques—iron metallurgy, for example—developed until it reached a level that was "coherent" with the state of the art, possibilities, and productivity of other related technologies. Thus, in the eighteenth century the new power of the steam engine, which offered, among others, the possibility of pumping water from mines, led to the development of coal extraction, which, in turn, led to a new form of iron metallurgy. The building of ships and trains from iron, and improvements in mechanical construction, were then possible, which fed back into the development of the steam engine, and so on. But before reaching a smooth working level, certain bottlenecks had to be removed. Upstream in the sequence, coke allowed the production of more cast iron, but the transformation of cast iron into iron was slow until 1783–84 when Cort developed puddlage, a technique by which coke could quickly be decarburized. Cort's work made it possible to speed up downsteam operations as well, by designing a steam-powered flatting-mill. Gille also presented an interesting sketch of the race between spinning and weaving in the eighteenth century, the efficiency of the one technique passing that of the other for a time, before being passed in turn, until an equilibrium was finally reached. It has yet to be explained why a system already in equilibrium should itself move toward a "more equilibrated" one. Gille's approach in terms of bottlenecks is nevertheless much like my own "strategic operations," to which I shall return later.

Gille's analysis of the dynamic relations between technological systems and other systems is a more or less intuitive one. Let us take, for instance, his analysis of what he called "blocked systems." He remarks that from the twelfth century to the present, technological revolutions in Western Europe have succeeded each other more or less regularly. All other technological systems appear to have come to a halt, at one time or another. This is true of the Greco-Roman empire, of the golden era of Egypt and Mesopotamia, of Pre-Columbian America, and of the modern-day Islamic world and China. Unfortunately, Gille has merely pointed to these blockages, he has not explained them, except to say that the reason had something to do with other components of the social and economic systems. But we have no details on these relations.

And this situation appears to be quite general when one looks at the work of other historians. Needham (1969, 1970), for example, attributes the long-lasting stagnation of modern Chinese technology to the non-development of scientific thought, bureaucracy, absence of capitalism, and the isolation of ancient China. For Rostow (1975), on the other hand, China lacked a spirit of invention. But what are the precise effects of bureaucracy, for instance, on the transformation of a technological system? What are the real channels of such influence? This type of question is not answered by historians; and archaeologists or ethnologists or historians who do tackle these questions might better approach these problems by dealing with a more limited range of technologies. Instead of merely suggesting that ideology has something to do with the general evolution of technological systems, anthropology should be able to address concretely the social representations of technologies, which in turn might lead to results of some interest for the study of both prehistoric societies and "great civilizations," not to speak of technological change in our own industrial societies, as we shall see.

In sum, technologies are social phenomena; they are composed of five basic elements related in a systemic way to each other and to other social phenomena. Let us now see how anthropology has in fact dealt with technology.

CLASSICAL ANTHROPOLOGICAL APPROACHES TO TECHNOLOGY

Curiously, between the end of the last century and World War II, our great anthropologist ancestors were very interested in what was then, and should still be, called "material culture." Boas, Kroeber, Haddon, and many others gathered thousands of objects and left at least that many pages of reports in the Smithsonian Institution, Bureau of American Ethnology, Field Museum of Natural History, Peabody Museum at Harvard, American Museum of Natural History, Pitt-Rivers Museum in Oxford, Museum of Anthropology in Cambridge, Museum für Volkerkunde, Musée de l'Homme, and so on. These reports and objects are indeed well-preserved treasures of information. And yet they are but a small portion of what is needed to undertake an anthropological study of material culture. One will almost never find in a museum or in our great ancestors' reports what is required to reconstruct operational sequences, which are the basic data of any social approach to technological systems. As a consequence, the comparison of operational sequences, which is the basic methodology for bringing out the differences to be explained, is even less of a possibility.

A history of ethnologists' decreasing interest in material culture has yet to be written. It is perhaps partly related to our differing perceptions of science (which is considered noble and worthy of study) and technology (which is felt to be too ordinary a matter to concern serious scholars) (see also Sigaut 1980). Moreover, an interest in the history of science, and the relative lack of interest in the history of technology, parallels the situation found in ethnology. B. Reynolds (1983) noted the irony that ethnologists often spent so much time collecting artifacts in the field and transporting them to museums, but never touched these artifacts again, not even to dust them. In other words, it seems the need to describe and gather at least some artifacts was part of the ethnologists' perception of what their fieldwork should be, but often after gathering them, they seemed at a loss as to what to do with them.

What was lacking in museums, and in the reports produced in the artifact-collecting days of ethnology, were data on the material worked by the artifacts and on gestures, not to speak of the knowledge involved in the use of these artifacts, or during the technological actions in which they were used. Even movies often show an esthetic bias. Although video tapes cannot record the social context or the tacit knowledge of the actors, they could easily give us nearly complete information about the physical action on the material; but this requires, at a minimum, that the people doing the filming be aware of how to describe action on matter (Esparragoza 1983).

Despite these problems, museums and, once again, the reports made by those scholars who brought most of the collections into these museums, contain what will forever be the only record of the technological systems of now extinct societies and civilizations. It is our task to extract from this huge store of information reliable data which will sustain comparisons. As archaeologists know, we can often reconstruct how artifacts were made and used from their shape, dynamic features, patterns of wear or physico-chemical composition (Swanson 1975). We may even experiment with some of these manufacturing processes and uses (Coles 1973; P. Reynolds 1978). We might also have an idea of the systemic relations among elements of some sets of artifacts in a given society. Proceeding from the features of artifacts to the gestures and materials, we might even enter the realm of the social representations of technological systems. But everything, or nearly everything, in this respect remains to be done.

Closely related, from a museological perspective, to the quest for artifacts is the ethnological study of costume, an interest which parallels the stylistic approach to artifacts such as stylistic studies of ceramics in archaeology, or the comparative ethnological studies aimed at statistically defining "cultures." While the study of the distribution of cultural traits is the most elaborate of the classical ethnological approaches to material culture, it leads to correlations that are, from an ethnological point of view, extremely difficult to interpret, a difficulty which stems from the geographic scale of the culture areas included and the significance given to various technological traits included in the survey (see Driver and Massey 1957 for an example).

I shall return to the ethnology of costume later, when I consider whether some technological traits function as symbols. For the present, let us keep in mind that this approach is often based on an informational perspective—the decoration or shape of artifacts (including costumes) display stylistic variations which inform those who look at the artifacts about some aspect of the social identity of the individuals who wear or use them. "Style," whether in ceramics or costume, is restricted to features which, in the great majority of cases, exercise no action on the material world (except that of being visible). This traditional approach to style, therefore, leaves aside the way people and society physically act upon their environment, how things are made, how a pot allows cooking to take place, how a piece of clothing covers or protects one's body from heat or cold. In other words, this approach operates as though *only* an immediate symbolic behavior were involved in a technological process.

There is another kind of approach to material culture that is more concerned with physical features, and which may be characterized as the search for a direct, one-to-one relation between technology and society. The examples that follow are taken from both historians and ethnologists.

The extended debate on the correlations between slavery and the lack of technological improvements in European antiquity is a case in point. For Lefebvre des Noëttes (1931), it was an improvement in the forces of production—in this instance, the development of the harness—that rendered slavery useless. Bloch (1935) was to show, on the contrary, that the decline of slavery came first. Other scholars have argued instead in favor of the primacy of people's social representations of their technologies or of technology, generally speaking. Finley (1965) demonstrated that the Greeks and Romans felt no need to improve the productivity of technological operations. For him, too, a decrease in available human labor came first and, by the end of the Greek and Roman empires, the principal factors were political ones—high bureaucratic and tax pressures, and deterioration of statuses. Schuhl (1969) thought that the Greeks and Romans had an "antimechanization" mentality, because the abundance of human labor did not generate a need for mechanization. Aymard (1959), in turn, argued that because slavery was socially acceptable in Greece, craftsmanship was both socially and intellectually belittled (as it also seems to be in today's ethnology). So, for these scholars who are among the very few to have given attention to the role of technology in history, opinion is divided between technology as the driving force behind important social transformations or ideological features as the basis of major technological improvements.

A dangerously simplistic explanation of the role of technology in history characterized Lynn White's (1962) approach to the rise of feudalism. For this author, the introduction of the stirrup into Europe in the eighth century (from the East) gave horsemen a better seat and thus enabled them to charge holding a spear. The development of heavy cavalry followed. Knights then became specialized horsemen having both great military and political power. In this line of reasoning, feudalism might merely be the result of introducing the stirrup! Fortunately, by carefully using an increasing number of detailed studies on the evolution of particular technologies, combined with a vastly improved overall level of available historical knowledge, historians now use far less simplistic and more systemic views of the relationships between technology and society. This is, of course, true of the pioneering work of Gille. It is also true of Sigaut's (1985) synthesis of the evolution of pre-industrial European agriculture. Braudel's *Afterthoughts on Material Civilization and Capitalism* (1977) also escapes any kind of reductionism.

Short-cuts can be found in ethnological arguments as well: for Watson (1965, 1977) and Sorenson (1972), the introduction of the sweet potato in New Guinea, some three centuries ago, attempts to explain no less than a rise in population, an increase in warfare, and the first steps toward social stratification. So we see there exist many sorts of materialist approaches to technological systems, some more simplistic than others.

ECONOMIC ANTHROPOLOGY AND TECHNOLOGICAL SYSTEMS: A ONE-SIDED APPROACH

Nor is recent economic anthropology free from proposing such direct relations between particular aspects of technology and huge segments of social organization. This is the case even among those Marxist anthropologists who should be cognizant of the crucial connections between forces of production and social relations of production, the former being the means (intellectual as well as physical) that societies use to extract their subsistence from their natural environment; the latter being the social relations which determine the production, circulation, and redistribution of material goods (Godelier 1977). For example, Terray (1972) focused his analysis of Meillassoux's (1964) Guro (Ivory Coast) data on those particular social relations of production which concern how the forces of production are put to use, and he considered only the division of labor, emphasizing types of cooperation: (1) "simple," when every participant is in charge of the same technological operation; and (2) "complex," when different operations are performed by the people who cooperate. He found direct correlations between these abstract features of technological processes and forms of social organization.

According to Terray (1972:137), complex cooperation

is "realized" in what we have called the tribal-village system. In the category of relations of production, ownership of the means of production is collective and the rules of distribution are egalitarian.... Simple cooperation reveals the presence of the second mode of production, which is "realized" in what we have called the lineage system.... In the category of relations of production, the means of production are owned collectively, but a single individual holds them on behalf of the group.... Finally, this mode of production entails authority functioning continuously, entrusted to persons selected by virtue of their age.

Meillassoux's (1967, 1981) comparison of the material basis of hunter-gatherer societies, on the one hand, and agriculturalist societies, on the other, is based on an even more abstract analysis of the forces of production. For him, land among hunters and gatherers is an "object of work" with no preparation involving work, whereas among agriculturalists it becomes a "means of work" in which work has previously been invested. Because of this lack of "investment," work has an immediate return in hunter-gatherer societies. Cooperation stops as soon as hunting stops, so that the sets of action are unstable. The game is immediately consumed; there is no delayed distribution of the product of hunting. For these reasons, according to Meillassoux (1967:101), "these features do not give a basis for the construction of a centralized and lasting political power." Among hunters and gatherers, the relative equality of men and women, and the diffuse (in Meillassoux's view) nature of kinship, are explained in the same manner.

Among agriculturalists, on the other hand, who exemplify the "domestic mode of production" (Meillassoux 1981:33–49), the social relations initiated during the agricultural process go beyond the limits and the moment of this process. Agriculture is a discontinuous process, because one needs to have seeds to put the land to use, and one has to have enough food to survive while waiting for the harvest of what has been planted. In other words, agriculture, being a long-term form of production, involves a cycle of advances and returns. For Meillassoux (1981:42), these abstract features of agriculture

create lifelong organic relations between members of the community; they support a hierarchical structure based on authority (or "age"); they constitute functional coherent economic and social cells which are organically linked through time; they define membership, as well as a structure and a power of management which fall to the eldest in the productive cycle.

However interesting Meillassoux's theory may be, it remains that his approach, like Terray's, gives no room to the study of the basic physical features of the forces of production. One wonders if Marx was really thinking of such one-way and indirect relations between productive forces and social relations of production.

Godelier is a third French Marxist anthropologist who has recently given some attention to forces of production, and he is the one who has gone the furthest into their physical organization. In his analysis of Turnbull's (1966) data on Mbuti Pygmies, Godelier (1977) started from the organization of the hunting process itself. Although he still left aside many physical aspects of the use of nets in hunting, not to mention bow and arrow and spear hunting (but see Bahuchet 1985 and Demesse 1978, 1980 for descriptions of these techniques), Godelier derived three "constraints" that he believes arise from the necessity of assembling hunting parties large enough to assure hunting success. These constraints (dispersion, cooperation, and fluidity) in turn affect Mbuti social organization: kinship rules, political relationships, and religious behavior. At least Godelier's analysis deals with the physical process of hunting. But, once again, not at length. For him, as for Terray and Meillassoux, the forces of production are a given, and a black box to boot. For these scholars, constraints result from the organization of the production and, for them, anthropology starts with the study of the effects of these constraints on other social phenomena.

This "constraints" approach might be an unspoken manifestation of the concept of "level of forces of production" which, except in general views such as Leslie White's (1959) or Mumford's (1934), is not much help in understanding the transformation and evolution of technological systems and societies, at least on the analytical level used by ethnographers.

Let me note in passing that Marx's own approach to the effects of the forces of production, from an evolutionist perspective on economic and social organizations, is far more sophisticated than the caricature sometimes given of his work. As Digard (1979) has shown, saying, for example, that the steam engine led to the development of industrial capitalism is the kind of pedagogical abridgement of Marx that has done the most damage to the study of the forces of production among Marxists themselves.

The anthropological approach to technologies, or forces of production, in terms of their efficiency and output (curiously rare among Marxist anthropologists, but see Godelier 1971, 1973), has been developed mainly by "cultural ecologists" such as Rappaport (1968) and Lee (1969, 1979). They have devoted much attention to the results of existing technologies among the Maring and the !Kung, respectively, but they did not study these technologies in themselves from an anthropological point of view. Once again, what is emphasized is the output or effects of the technological system on other social phenomena, or the reciprocal effects of some social features (the ritual cycle in Rappaport's work) and the technologies. As useful and necessary as they are, minute measurements of the efficiency of technologies (agricultural and stock-raising technologies in the Maring case) do not constitute an adequate anthropological account of technological systems as *social productions*.

A COMPLEMENTARY APPROACH

It is noteworthy that Marxist and cultural ecological approaches could profitably be joined together: a study of the *effects* of a technological system on a society should consider simultaneously the social relations of production which correspond to a given set of forces of production, as well as output and efficiency. Both of these approaches are essential in any comprehensive study in economic anthropology, and they are a necessary stage in an analysis of the relationships between technology and society. Nevertheless, both fail to consider the social dimension of physical actions of technologies on the material world.

As I have already emphasized, technology is a social phenomenon and exhibits many systemic aspects; anthropological study restricted to the mere effects of technology on society therefore does not suffice. Technologies must be considered in a general anthropological perspective as social productions that are determined by, or better, are compatible with, other social phenomena. Because the features of these technological systems are not the simple result of physical constraints, either constraints internal to the technologies themselves, or constraints arising from the natural environment, the question of the influence of social choices has to be seriously raised.

Voluntarily or, more often, unintentionally, societies accept or ignore technological answers that they might either develop themselves or borrow from other societies. Following Lévi-Strauss (1976:11), I shall speak of technological *choices* in this context. Except for a very few instances (e.g., the choice of nuclear power; blocking adoption of the supersonic transport), these choices are not the result of documented individual or collective decisions. Rather, it is as if, during its history, a society, for unknown reasons, had come to rely on one particular technique, even though others were potentially available to it that could have produced the same kind, or nearly the same kind, of result. It is this open possibility of developing two or more alternative techniques at a given time in a society's history which leads me to use the term "choice."

Ethnology and archaeology have been interested in such choices for a long time, particularly in those which are related to the informational dimension of technologies, as I have already noted. The functions of some technological features which carry "signs" or "symbols" have been investigated: decorations on pots, so-called "nonfunctional" details of lithic industries, shapes of kettles, parts of costumes, and so forth. Less often, scholars have been interested in the social context in which these signs or symbols are produced. This, for example, is the case in Hodder's (1982) *Symbols in Action* or in Wiessner's (1984) psycho-sociological theory of headband production among the !Kung San. It is important to realize that the traits on which these studies usually focus are but a limited—very limited indeed—part of technological systems; and that to consider only these particular stylistic traits results in nothing less than leaving aside the most material, or physical, aspects of the social action on the material world. I submit that anthropology must also investigate the social production of other technological features, the function of which is more physical (dealing with matter and energy) than informational. The purpose of an anthropology of technological systems could then be to investigate whether some technological choices are arbitrary from a *technological* point of view (they of course will not be arbitrary as social productions). If such choices, independent of any physical necessity, do exist, it is important to understand how they are socially produced, and to what extent these choices influence transformations of technological systems and societies.

Most scholars no doubt will admit that a design on a potsherd, the color of a sweatshirt, or the shape of a tile may result from social decisions that have little to do with the efficacy of the pot as a container, the body protection afforded by the sweatshirt, or the waterproofing qualities of the tile. But I wonder if most anthropologists recognize that a physical action on matter can result from, or be modified by, choices, the logic of which is not strictly "technological"; that is, the "reason" for which a particular choice has been made is, above all, related to phenomena other than physical ones.

At least we know that such choices do exist. We need only consider technological processes that fail to fulfill the physical action on matter for which they were designed, or which result in the misuse of technological devices supposed to make one's life easier, or even assure one's survival. For example, some people fight without shields against opponents who use them (Brown 1910:161); others do not copy the more effective arrows used against them by their enemies (see below); still others jeopardize the survival of hundreds of soldiers by supplying them with improper ammunition for their rifles (Fallows 1985). Even in the most advanced domains of modern technology, as in the designing of nuclear missiles (Armacost 1985) or planes (see below), can one see technical options of engineers that are clearly being influenced by representations, beliefs, and ideas which have little to do with basic scientific, technological, or even economic logic.

All of these technological actions—failures in most of these examples—result from nontechnologically based choices. How much influence do such choices have on technological systems? Before answering this question, or rather, before trying to set out a program which will allow us to answer it, we have to specify how to identify such choices.

Ethnology is a study of differences, and archaeology is in turn dependent on ethnological analogies. The anthropology of kinship and marriage, "symbolic" anthropology, and "political" anthropology all rely on detailed data gathered over long periods of fieldwork and analyzed in thousands of articles and books. It is an inevitable conclusion that an anthropology of technological systems, too, can only be developed if it has access to precise, detailed, and, above all, comparable data. A study of the relations between technology and society must necessarily start from the study of differences, of variations in technological actions, observed in one particular society as well as among many societies, through time and space. Without such studies, the study of technology is just a black box which remains outside of anthropology. To study these variations, it is necessary to "see" them, whether they concern differences in the material used, in the means of work, in the gestures and operational processes, or in the specific knowledge involved. I shall return later (Chapter 2) to the means available for discerning differences in technological actions, and especially to the problems of scientifically describing and classifying technological traits. Obviously, these classifications are ours, and should not be confused with indigenous classifications of the same technological traits.

Let us now consider the kind of immediate results that may be obtained simply by pointing to coarse variations in technological processes. This exercise will illustrate how technological variability can inform us about nontechnological phenomena.

To look for differences (variations) is to take into consideration the discontinuities in particular technological traits of material culture, both through time and over space. For example, a spear-thrower may have three notches in its handle instead of two; married men may wear red vests, not yellow ones; the sap of *Rhus taitensis* may be spread on tree branches to catch birds, but may not be used as glue to strengthen the bindings of arrows; fields may be weeded with a machete or by hand rather than a hoe. The second step of the approach is to ask if these discontinuities are related directly to physical phenomena, or if their "meaning" or explanation lies outside the domain of an action on the material world; that is, in social phenomena of another kind.

Variation here refers not only to the presence or absence of a given technological feature in a particular human group, but also to the many ways that exist within one society of performing one particular technological operation, with roughly the same result. Given such variations, the first question to ask is if the way the operation was conducted can be explained by some technological-physical necessity. If not, the social, or nontechnological, context of these variations has to be explored. For instance, when considering the salt-making processes used in salt marshes along the Atlantic coast of France (Lemonnier 1980), observation shows that in Guerande, north of the Loire River, people gather the "fine-salt" which forms just on the surface of the brine covering the crystallizing areas. Some miles away, but south of the Loire, in Vendée, the salt-makers do not collect this product and declare that, should they do so, the "coarse-salt," which crystallizes on the bottom under the brine would not "grow." In order to make it "grow," they break the layer of fine-salt and redissolve it to "feed" the crystals of coarse-salt. Now it is true that dissolving the fine-salt adds some salinity to the brine; but this often slows down the crystallizing process instead of speeding it up; and, more important, it is not true that removing the fine-salt prevents coarse-salt from crystallizing, as shown in all the salt marshes north of the Loire. So this variation in the salt-making process remains incomprehensible from a purely physical point of view.

Let us now turn to the social context of the production and distribution of fine-salt. South of the Loire, this by-product has always been of no economic value, because it had no economic outlet. North of the Loire, on the other hand, it was harvested by *porteuses* (female carriers) who carried the daily harvest of coarse-salt on their heads from the crystallizing areas to the outer dike of the salt pond. These women came from outside the family of the cultivator of the salt pond, and were often from poor families. The fine-salt was their only salary for carrying the coarse-salt, which, in the nineteenth century, was bought by the fish canneries of the nearby harbor of La Turballe. When these canneries closed their doors in the 1930s, the *porteuses* disappeared, having no more economic outlet for the fine-salt. The cultivator's family retained the habit of harvesting part of the fine-salt for their own consumption (and after 1975, for sale in the new natural foods market).

In the French salt-making example, explanation of the observed technological variations clearly lies outside the domain of strictly technological processes, and instead is tied to the distribution of one of its by-products. But explanations of technological variations are not always so easy to find. As we shall see in the case of the Anga of Highland New Guinea (Chapter 3), the covariation of technological features which have no functional link to each other cannot be explained by any physical or environmental necessity, nor can the covariation be explained in terms of the more immediately obvious social context in which these technological features were produced. Instead, the explanation must lie somewhere within the social representations of these technological features, the locus and logic of which remain to be explored.

Generally speaking, variations in any of the five elements of a technique provide a starting point for an anthropological investigation of technologies. They are the phenomena to be explained. The previous example of salt-making has shown that sociological explanations of technologies can be found which have nothing to do with style, and which are unrelated, or at best only partly related, to physical actions on the material world.

Particularly interesting are those variations which concern what I call "strategic operations." These are operations that cannot be (1) delayed, (2) cancelled, or (3) replaced without jeopardizing the whole process or its final result. It may seem that all the operations in a technological sequence are equally necessary. This is not the case, however, as some examples will show.

Operations That Cannot Be Delayed

When using cyanolicric glue, one has to clamp both pieces being glued, at the right time, and for at least a minimum duration, otherwise the pieces will not stick. Nor will the result be very good if one forgets to remove a frying-pan from the fire when the steak is cooked. And in the same category of phenomena is the fact that after opening to full throttle, it takes 17 seconds for a Boeing 727 to start accelerating. Pilots must remember this in the event of an interrupted landing procedure. These examples show that there are nonreversible states of matter which require performing a certain operation precisely when it is physically needed, not before, not after, but right on time.

Operations That Cannot Be Cancelled

After reaching the "S1" speed (290 km per hour for a 365-ton Boeing 747), the plane has to take off, whatever trouble may appear aboard, simply because it can no longer stop without great damage, and it is safer to try to fly it. Many examples can be found of the devastating results obtained if one cancels particular technological operations: if the water taps are not turned off at the appropriate time, the bathtub runs over; if the driver of a car does not turn the steering wheel when rounding the bend, the car leaves the road.

Operations That Cannot Be Replaced

Many elements involved in a technological process can be replaced by others which fit, more or less. If you do not have Bordeaux wine to make a sauce, for instance, you may try a Burgundy. If you have to make vegetal salt and do not find *Coïx* spp. canes in the natural environment, you can use Impatiens (family Balsaminae) or ferntrees instead. But sometimes an operation (or any other element of the process) cannot be replaced; that is, there are no alternatives to that operation. For instance, if it has been raining so heavily that the ground has turned to mud a few days before the grain harvest, there is no way to use a combine, which would sink into the mud, and the whole grain crop will over-ripen and spoil. If, for some reason, no more Korean-made sheet-metal came to Detroit, factories would stop producing cars for awhile. Similarly, some peoples of New Guinea would have had difficulties in gardening if the trade routes through which they obtained their stone tools were closed, because of warfare, for instance. The right quality of stone could not be found in their own natural environment, and a shortage of stones would have made agriculture difficult (Godelier 1973). This is an illustration of how a unique bottleneck can endanger a major component of an economic system.

These are obvious examples of "strategic operations" or strategic elements of a technological system; but in most cases only accurate descriptions and analyses of operational sequences can lead to the identification of potentially critical bottlenecks. I believe it would be easy, and worthwhile, to investigate the bridges that link particular technological and social phenomena, especially any social control exerted by a given group over strategic operations. For example, the socioeconomic context of the development of gears in the Middle Ages, or of the patents and financial control in the nineteenth century that influenced the invention of a device which allowed internal lubrication of a drill in operation, could all be investigated. The former (gearing) was responsible for the growth of mills and derived machines, and the latter made it possible to speed up the making of hubs, which, in turn, led to the development of the bicycle and automobile industries (Rosenberg 1963).

It might be of some importance for economic historians to investigate such immediate links between technological and socioeconomic systems. In this respect, Godelier's (1986) suggestion is interesting: that relations of domination and exploitation could be seen, at a given time, as an exchange and, more precisely, as an exchange of services. This possibility of linking the rise of social inequalities to a former control of real, or imaginary, strategic operations might be of some utility. At any rate, this provides a new, and more focused, perspective on the study of the question of specialization and the rise of social differentiation.

One of the best-known ethnological monographs provides us with an example of the kind of investigation that could be made in this direction. In the Trobriand Islands, the magician, according to Malinowski (1935:78–79), was a great gardener as well as a good meteorologist. He had the choice of performing either a short or a long ritual before the start of gardening operations. Following Malinowski (1935:108), we can suggest that the magician may have performed the short ritual when he believed that agricultural activities should begin soon, and the long ritual when he believed that they should be delayed. One wonders, therefore, if this was not an effective way of adapting the agricultural cycle to the weather conditions, and thus if his position had something to do with his particular technological knowledge. At the same time, it is of course the ethnologist's job to understand why in many societies no particular power has emerged from the control of given strategic operations, or why some strategic operations escape any particular social control.

Up to this point we have at least raised some new questions, which ultimately can lead us to the recognition of direct relations between technology and society. Those relations will originate in the physical actions themselves, and will be far less simplistic than the kind of stirrup-to-feudalism ones discussed earlier. It has to be remembered that, on the whole, the anthropology of technological systems does not deal with such direct relationships, but with more subtle or hidden ones, those which through a society's social representations of technologies influence the physical action on the material world.

In the end, comparison is the key-word. All of the examples given above rely on investigation of the variability in technological processes or of the social context (or control) of these technological processes. Now that we have made a general survey of the kind of questions that may be asked in an anthropology of technological systems, we must go back to where we started, to the classifications of technological data necessary for comparison. But before classifying, it is necessary to know what to classify, and, before that, we must construct the data. This is accomplished by recording operational sequences.

Chapter 2

From Field to Files

Description and Analysis of Technical Phenomena

OPERATIONAL SEQUENCES: THE BASIC DATA

The purpose of fieldwork in the anthropology of technological systems is to gather and construct comparable data concerning socialized actions on the material world. Whether dealing with a shamanistic cure, a pig killing ceremony, or a cockfight, an ethnographic study ideally should eventually embrace, step by step, the whole society in which the ethnographer is working. For purely material reasons (brevity of the ethnographer's life-time; duration of research grants), this is not possible. It is not practical either, for it must be kept in mind that the data we gather have to be not only useable, but used, and now. For this very reason, the question to ask before starting "technographic" work is not "what can I describe" but, rather, "what data do I—or any ethnologist—need, among those data which are available to me?" The answer is: operational sequences, the details of which can be compared to details of the same kind in other operational sequences. This narrows the field of our investigations considerably.

First used by Leroi-Gourhan in his lectures in the 1950s, the concept of "operational sequence" (*chaîne opératoire*) has been defined as "a series of operations which brings a raw material from a natural state to a manufactured state" (Cresswell 1976:6). As there are many techniques which do not lead to the making of a product, I consider that an operational sequence is more simply the series of operations involved in any transformation of matter (including our own body) by human beings.

It is possible to give an extensive list of the kinds of phenomena and information that one can record in the observation and comprehensive description of an operational sequence. According to the CNRS (Centre National de la Recherche Scientifique) research team's "Techniques et Culture" (1977), it could, and very often should, include information on the place, date, time, and duration of the action; the people acting; the matter being worked and its successive states; the tools successively used and the movements made while using these tools, with details on the kind of percussions and prehensions involved; the division of the process into steps, including both divisions made by the informants themselves and those made by the ethnographer; a description of the tools, parts of tools, and their names. An operational sequence might further include such things as meteorological conditions during the technological operation; a description of the clan or lineage that did not participate, but gathered where the action was being performed; an explanation of why the main informant was sulky or why the alarm clock did not wake the ethnographer up on time.

The data on technological systems, generally speaking, must also include an inventory of the possibilities offered by the natural environment of the given society in which the ethnologist works. These include the potentially useable plants, animals, "plastic" materials (such as water or gas), raw materials, and forces (water or wind). Yet another inventory would be the plants, animals or materials that are actually used by the society. As already noted, the mere comparison between the possibilities which are available to people and those elements which are actually used yields critical information on the technological knowledge or social representation of technologies shared by the members of this society. The ethnographer is, of course, allowed to ask for help from botanists, zoologists, geologists, and other specialists when his or her own scholarship does not fit the problems encountered.

How people talk about the technological actions performed in their society is another fundamental domain of investigation for the ethnographer. Comparison of the inventories of possibilities available to and actually used by a society, in conjunction with a linguistic approach to the society's technological knowledge, would lead to a kind of ethnoscientific study of material culture, or "ethno-technology." I shall come back to this point in Chapter 4.

It is one thing to know what might (or should) be described for the purpose of an ethnology of technological systems, and another thing to know how to observe, describe, transcribe, and analyze the operational sequences with which we arbitrarily sum up these technologies. The following discussion provides a brief commentary on these basic steps in collecting and recording the data essential to any ethnographic study of technology.

Observation

It is not my purpose here to explain the notion of "participant observation." Moreover, the would-be "technographer" should be forewarned that it is often very difficult, and at times disastrous, for the observer to try to perform the same technological actions that he or she is describing. This involvement may well allow the ethnographer to gain a better understanding of what is going on, but if this entails breaking things, boring the informants, or seriously interrupting a crucial activity, it might be better not to try to be a potter, a hunter, a salt-maker, or whatever. Furthermore, it may be difficult to become a reasonably good potter or hunter or salt-maker without a long apprenticeship. And finally, being able to do the thing "from the inside" might give us the dangerous illusion that the ethnologist can actually become the craftsperson. The best solution is to complement the information being gathered by relying as much as possible on actually performing the technological actions to be described, all the while taking great care not to project a personal ethnocentric perception of the task at hand. In 1935, Lucien Febvre, in a well-known issue of *Annales d'Histoire Économique et Sociale* devoted to technology, called for a "technological history of technologies." This means that historians (or ethnologists, or archaeologists) must be aware of the physical phenomena taking place before them; but this does not necessarily mean that they have to be able to perform the kinds of actions they are describing. There may be no way of participating in cassowary trapping, or grass skirt-making. And yet both of these actions have to be described.

It is fortunate for the ethnographer that technologies are often an area in which actions and behaviors are repetitive. Ethnographers thus have the possibility of training themselves to look at and understand what is taking place (which often is far from obvious). It may take days to understand a knotting technique or a twist of the fingers.

These are cases in which it is valuable to try to do the things oneself because a little familiarity can help one describe a process. But this should not be confused with mastery of a craft. On the other hand, it is always risky to put off until tomorrow what one could have observed or described today. Some technological actions never occur twice during a field-season, and waiting for another opportunity to observe a missing link in a technological process may take as much time as finding a missing link in a myth or ritual. Chance plays an important role here, and deciding whether to observe a given technological action today or tomorrow is all a part of the glorious uncertainty of ethnography. The only certainty is that some techniques inevitably will be used when the ethnographer has better things to do than describe them. This is not unique to the ethnography of techniques, but many ethnographers have given us poor accounts of technologies simply because their descriptions were interrupted by what they felt at the time were more important investigations. Many details may easily escape attention if one is not aware of their importance for an ethnology of technology. For instance, many time charts are wrong because the ethnographer forgot such secondary technological actions as bringing in raw materials. An ethnologist normally would not be able to observe a shamanistic cure *and* draw genealogies at the same time. It is therefore important to remember that describing technological actions is also a full-time field occupation.

Description

So, here you are ready to observe the making of popcorn in a middle-class American kitchen. Now, by what means are you to make the description as comprehensible as possible? Our great anthropologist ancestors made drawings. This gave them time to understand what they were drawing (similarly, it is well known that important aspects of botany can be learned best by drawing plants). Ethnographers can take pictures as well. Some even make drawings by tracing them from pictures. So, a still-camera, a pencil, and a watch are mandatory equipment. With a little training, it is possible to watch, write, take pictures, and note the time without interrupting the people at work. This does not mean that they may not be asked to slow down, to stop, or to repeat a given action that the observer missed. The time chart only has to be modified accordingly.

The would-be "technographer" should note that, for unknown reasons, we have a tendency to snap a picture at the same instant for a given technique, resulting in many pictures of the same movement, instead of a sequence of photos covering the full range of movements involved. The use of a motor-driven camera eliminates this difficulty, as the camera automatically cuts the process into regular segments. Another trick is to talk into a tape-recorder whenever the action being described is too fast to be written down at the same time, but still slow enough for dictation. The noise of the camera "clicks" on tape may even help identify the action taking place in specific pictures, as well as providing a measure of the duration of each action.

Here we reach the practical limits of describing the material aspects of a physical action on the material world. But using a movie or video camera can easily extend them. In these matters, everything is only a question of time, money, and feasibility.

Transcription

Ideally, the use of a video camera by an ethnographer aware of what is going on should make it possible to obtain a comprehensive record of a given technological action. Awarenesss means that the ethnographer, for example, already knows what actions deserve close-ups, which also means that he or she has already analyzed what is going on—somewhat of a paradox. Transcription is dogged by another difficulty as well. Technological processes, as is the case with many social behaviors, are characterized by everything but linear sets of actions (cf. Chapter 1). Yet writing is distinctly linear, and, as a result, the very few ethnologists interested in technologies are still looking for an adequate means of graphic transcription of the phenomena they study.

There are two basic rules to abide by in this matter. First, it is not a good idea to develop too sophisticated a means of transcription before working with real data. It is true that nobody should go into the field without some theory in mind. But it may prove counterproductive to force technological phenomena into predetermined categories. Only a real encounter with butchering technology, for instance, will establish how many different categories are really needed to describe the way an animal is cut up. Thus, although it is desirable for all future specialists in, say, the ethnology of butchering to use the same vocabulary and categories, this cannot be done as yet. There are even a few scholars who, by continually trying to improve their techniques of description or transcription, end up never going into the field.

A second general but good rule is that graphic transcriptions should improve and simplify our understanding of reality, not complicate it. Figure 1 indicates which elements should be included in a basic description of an operational sequence. It is worth noting that the information on "social relations" may prove to be endless. What is stressed here is that sociological information regarding the actors and context of the technical process taking place also have to be recorded on the spot.

Transcription presents a more general problem, in addition to the specific ones above: how to classify technological features (a "feature" being anything from a car, to a stone adze, to a wrist movement). Rather than reviewing the numerous more or less appropriate means for classifying such features that have been used in the fields of human engineering (Chapanis 1965) and ethnology (Leroi-Gourhan 1943; Koechlin and Matras 1971; Lemonnier 1976), I shall just say a few words on the use of tree diagrams. By allowing the representation of simultaneous actions, as well as alternative actions, tree diagrams are well suited to the transcription of technological processes. By using colors and footnotes, one can make as many comments as one wishes. Keeping the diagram readable is another story. The general organization of such a tree diagram is shown in Figure 2. Each set of branches can be commented upon in a separate table of the type shown in Table 1.

Step by step, it is thus possible, at least theoretically, to display a number of technological processes at the same time. All that is needed is a room and a table big enough to lay the diagrams side by side so they can be related to each other, and the ability to read many tree diagrams simultaneously. From a practical point of view, the description of one technique may already cover many square meters of table space; the description of the salt-making process of the Baruya of New Guinea, for example, took me 85 pages of notes and 304 pictures (Lemonnier 1984a).

Analysis

It is by comparing transcriptions of operational sequences that an anthropology of technological systems will take shape. This, in turn, requires the use of classificatory criteria and categories of the same type, as we shall see. By superimposing, at least mentally, two or more tree diagrams which correspond to the operational sequences involved in a similar type of technological action in different societies (or at different periods of time in the same society), it is possible to display variations: any variation in the organization of the diagram corresponds, by definition, to a variation in some element of the operational sequences. These over-simplified transcriptions of the use of a hammer and nail by two carpenters from two different societies (Fig. 3) show that carpenter A repeats two operations while car-

From Field to Files

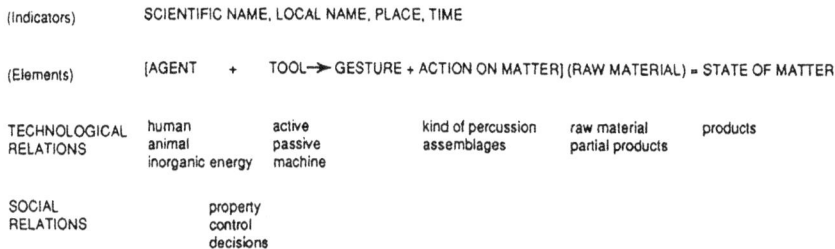

EACH STEP IN A SEQUENCE COMPRISES:

(Indicators)	SCIENTIFIC NAME, LOCAL NAME, PLACE, TIME
(Elements)	[AGENT + TOOL→ GESTURE + ACTION ON MATTER] (RAW MATERIAL) = STATE OF MATTER

TECHNOLOGICAL RELATIONS	human animal inorganic energy	active passive machine	kind of percussion assemblages	raw material partial products	products

SOCIAL RELATIONS		property control decisions			

Figure 1. Operational sequence ("work chain"): the basic model (after Cresswell 1983).

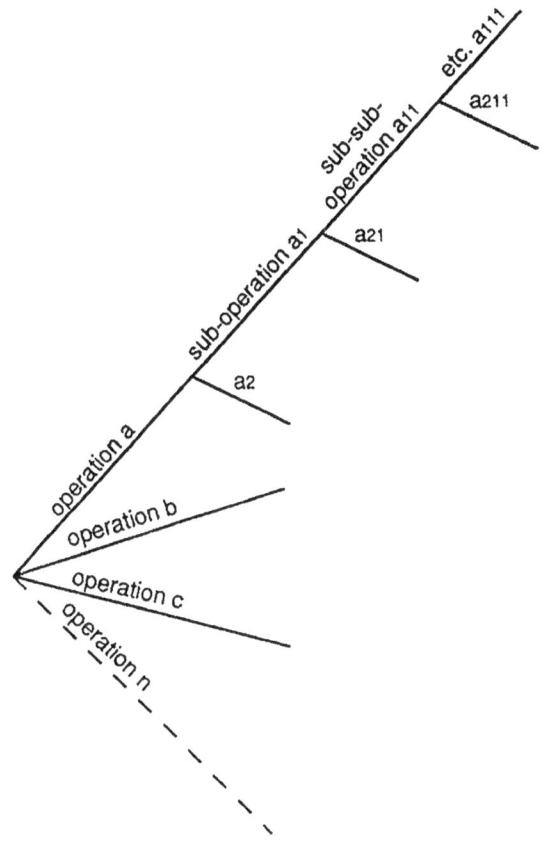

Figure 2. A tree diagram in which each set of branches represents the details of the operation immediately preceding.

TABLE 1
Comments on a Tree Diagram

	Operations			
	a(1)	a(2)	a(3)	a(n)
Short description of operation	-	-	-	-
Indigenous name	-	-	-	-
Time constraints	-	-	-	-
Other factors	-	-	-	-

penter B does not. Figure 4 presents an abridged (imaginary) description of the making of a pebble chopper (Fig. 4a) and details the analysis of this technical process along three axes: the first axis details each operation (in this example, operation 4); the second axis deals with the amount of energy involved; and the third axis provides a graphic synthesis of the whole operational sequence (Fig. 4b).

Furthermore, reading a tree diagram such as the one shown below in Figure 5 can tell us something about possible strategic operations: for example, operations a, b, and c must obviously be done before operation d can take place, and if c is the longest operation (c is the "critical path" in operational research), it is the one to be controlled with the most care. But a, b, and c must all be completed before the sequence can proceed to d. Operation e perhaps can be by-passed, if f and g can take place instead, and so on.

Tree diagrams may also show invariable subsequences, made up of the same operations, but which might appear in different operational sequences, possibly in diverse technologies. The search for such regularities is basic to an anthropology of technological systems: how are these units produced and used? What makes up their social representations? Is there any tendency to use a particular unit in many technological processes when it already exists in two or more processes? Is there a correlation between the efficiency of such units and the multiplicity of their uses? The isolation and definition of such units is related to the more general question, to which we now turn, of the classification of technological features.

WHAT AND HOW TO CLASSIFY

Anthropology has to compare in order to study the differences on which it comments. Comparison should be effected using classificatory criteria which are shared by as many scholars as possible. This

From Field to Files 33

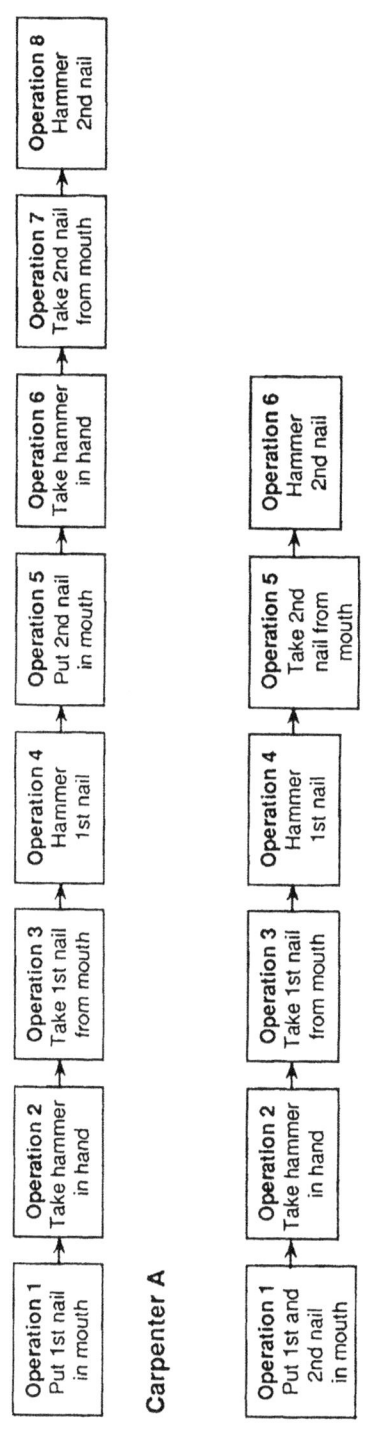

Figure 3. Two simplified operational sequences.

Elements for an Anthropology of Technology

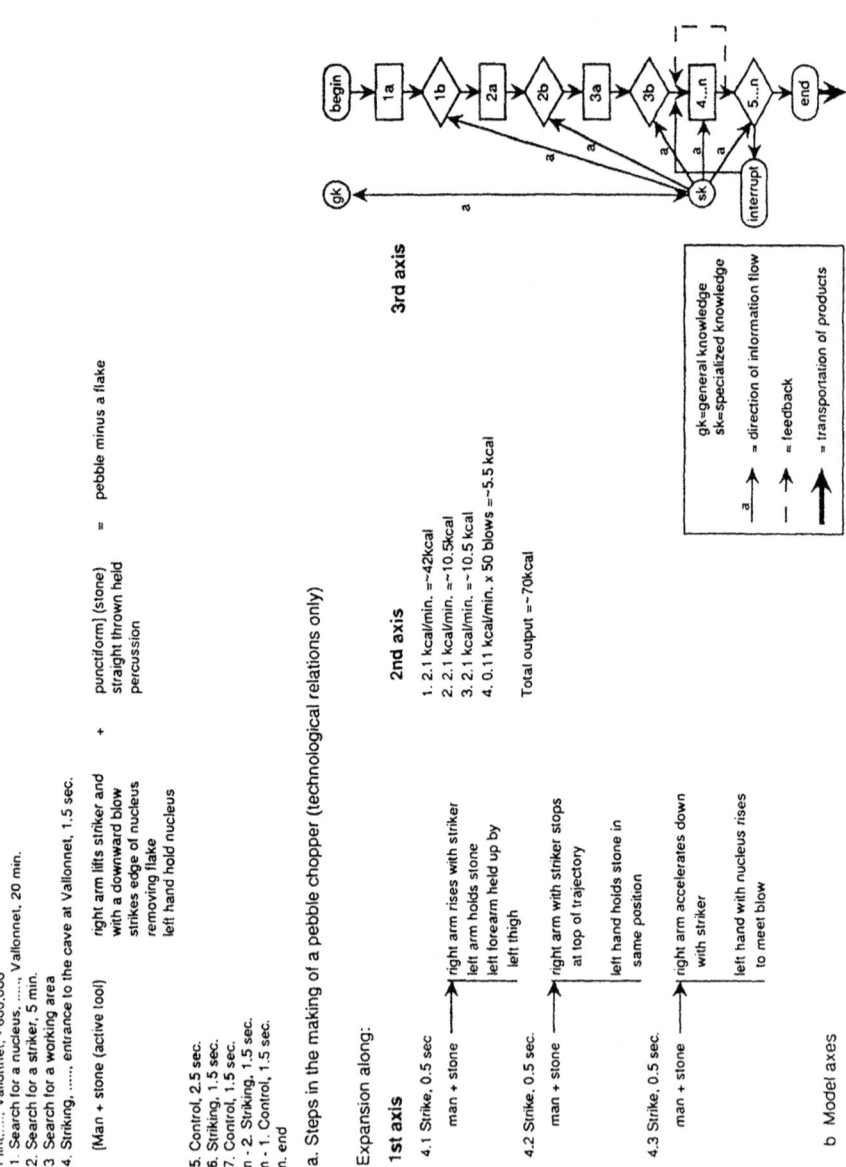

Figure 4. Description and analysis of an operational sequence. *a*, steps in making a pebble chopper (technological relations only); *b*, model axes (after Cresswell 1983).

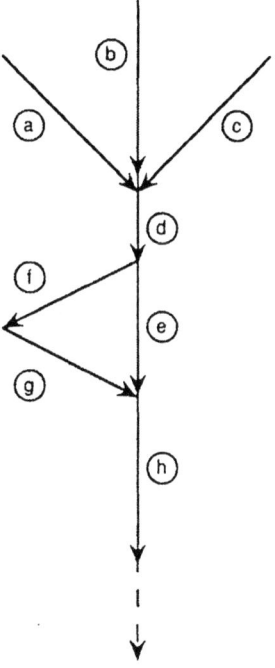

Figure 5. Diagram of an operational sequence.

is basic to many scientific approaches: to say that a rabbit is not a duck, or that a duck and a hen have more in common than a hen and a rabbit, the observer has to confront his own perception of these animals with a preexisting classification of animals, whether explicit or not. To say that a couch, a stone adze, and a snare differ, respectively, from an armchair, an axe, and a dead-fall trap also requires the use of classifications. But, whereas comparing the couch and the armchair may seem obvious, identifying a stone as being a part of an adze or part of an axe is less obvious. Moreover, there is a tendency to perceive each artifact of a given pair (couch/armchair; adze/axe) as resembling the other, whereas it is the differences that will be first noticed when comparing a snare and a dead-fall trap. It is possible, nevertheless, to look for similarities between snares and dead-fall traps. These might be found in particular parts of the two kinds of traps (the trigger mechanism, for instance), as well as in some of the technological principles involved in each case (e.g., tension and release).

These examples show that there is not one single type of classification, but many ways to classify actions on the material world. The multiplicity of possible classifications, if physical action is to be equated with a particular artifact, becomes even greater if by "physical action" one means "a whole technique" (e.g., making pots, building houses, or carrying babies in a particular society). In this case, it is operational sequences (not just mere artifacts) that will have to be classified. Thus, the elements involved in a given operational sequence will have to be compared to those found in another similar operational sequence (e.g., fence-building among the Baruya of New Guinea can be compared to garden fencing among other New Guinea tribes). The classification used will have to encompass very different sorts of elements: artifacts, movements, materials, representations of what is going on, number of people involved, duration, number of steps, and so on. In any case, classifications of technological traits must not rely only on their physical aspects, but also on the way they are made and used.

We shall leave aside the problem of comparing representations to focus on what seems easier: comparing (and thus classifying) artifacts and the gestures used to move them. Following Haudricourt (1968), we shall distinguish between two common types of classification: artificial and natural. Introduced by Linnaeus in botany, artificial classifications

> consist in using the same characters (number of stamens, for instance) in order to obtain the most inclusive possible classification. When no more can be extracted from the use of one criterion, another is chosen—for instance, the number of carpels in Linnaeus' classification—to continue subdividing the partitions already obtained with the preceding character. This is what is called an *a priori* subordination of characters.
>
> By the end of the eighteenth century, a time when theories on the evolution of living things had not yet triumphed, some botanists, unsatisfied with Linnaeus's method, suggested a more empirical and intuitive way of classifying, which they called natural classification. In this method there is no *a priori* subordination; in each case the importance of each differentiating character is estimated without preconceived ideas. Once evolutionary theory was accepted, scholars realized that natural classifications of living things showed their genealogical tree. [Haudricourt 1968:803, translated by the author]

Haudricourt once noted in a seminar that, strangely enough, artificial classifications are often based on criteria reflecting aspects of artifacts that "are of no use" in their physical action. It seems to me that, unfortunately, most of the "stylistic" classifications used in archaeology, as well as in museums, conform well to this observation.

If material culture is seen as being concerned primarily with physical action on the material world, the classificatory criteria to be used are, generally speaking, functional (or dynamic) ones; that is, related to the way the objects act on matter—in other words, how they work. Artificial classifications will be used only at the beginning of the research, to provide a rough order for the phenomena which are to be examined. In the second stage of the research, one then chooses the particular criteria which seem most pertinent to the analysis. Haudricourt (1968:804–7) gives an example of these stages; the geographer artificially classifies carriages as two- or four-wheeled, which provides an indication of their geographical distribution. But when examining the question from a historical point of view, the dynamics of carriages—how they are drawn, for instance—might also be considered. Then the distinction between carriages using shafts and those using poles may seem more relevant because it will be seen that two-wheeled carriages pulled by a pole preceded two-wheeled carriages using shafts by many millennia.

Therefore, while no single classification of technologies, or elements of technologies, will suffice, the number is finite. A systemic approach to an artifact or a given technological action offers a heuristic means of reducing the number of possible relevant classifications. According to Quilici-Pacaud (1987), who advocates such an approach, an *organic* point of view leads to the question: what are the components of the artifact or technological action? The answer takes the form of nouns: spring, wheel, counterweight, handle, blade, binding, screw, oven, combine. A *functional* point of view asks what is the function of this particular component? The answer to the functional question commonly is a verb: to store energy, to sustain, to roll and drive, to balance, to hold, to cut, to assemble, to heat, to harvest. The *relational* point of view asks how the components (or functions) are related to each other. The answer takes the form of diagrams showing the logical or physical links between components or functions.[1] Thus, although far from being infinite in number, there remain many criteria that can be combined in a classification. Fortunately, these criteria can, and must, be ordered according to their relevance for the study of a particular action on the material world, relevance being measured, for instance, by their ability to discriminate ethnologically significant differences or similarities (see below).

The graphic representation of these classifications may also vary. Their basic common purpose is to show in what respect two techniques or elements of a technique are similar and dissimilar. There are three kinds of such graphics: maps, tables, and tree diagrams.

Maps are an easy way to show the spatial presence/absence or distribution of any technological trait in a particular sociospatial entity: households in a village, villages in a county, tribes, language groups, and so forth (Bromberger et al. 1982–83). Once again, "trait" here could refer to any kind of technological feature: use of a particular machine, use of a specific design pattern, internal organization of operational sequences (e.g., linear, repetitive), or the evidence of the application of a given technological "principle." Figure 6 shows a distribution map of spinning technologies in Gîlan and Eastern Azarbaijan (Iran); Figure 7 gives the distribution of house types in Tuscany (Italy). The classification is implicit in a map inasmuch as the types being mapped are considered to be different. In the map in Figure 8, for instance, a "crude conical tipi" differs from a "Plains tipi." Maps can refer to progressively more detailed features of a technological phenomenon. Thus, the map in Figure 9 specifies particular aspects of the foundations (number of poles) of a particular type of house extracted from the map in Figure 8; and the map in Figure 10 provides information on a particular aspect of the social organization which appears in the operational sequence "building a house."[2]

Tables, too, show features of technologies according to the presence or absence of particular traits (see, for example, Table 2). They allow the simultaneous use of several criteria, although these remain limited due to the two-dimensional nature of paper. Table 3 illustrates the basic forms of agriculture on a worldwide scale. Here again—and in the making of tree diagrams as well—a common understanding of the categories used in the classification is implicit: "hoe" or "swing-plough" are assumed to be well-defined artifacts. Table 4 shows a classification of Dani (New Guinea) arrows developed by Heider (1970). Leroi-Gourhan's (1943, 1945) inventory of "simple" technologies was based on a classification of "elementary actions on matter": namely, prehension, percussion, fire, water, air, motive power, and transmission (1943:13–113). Table 5, another illustration of the use of tables, shows various actions and defines different types of percussion. For example, percussion #10 in the table is based on five criteria: perpendicular, linear, resting, transversal, with striker. Reference to these elementary actions on matter is made throughout the pages of *L'Homme et la Matière* (1943) and *Milieu et Techniques* (1945). In addition, maps and tables can be substituted for one another. For example, one can easily imagine the table that would correspond to Figure 8, in which tribes would be listed in the rows and each of the eighteen house types would be listed in the columns.

Tree diagrams are also partially equivalent to maps and tables. Their basic principle is that the nearer the trunk, the more the phenomena classified have in common, whereas the nearer the branches, the more they differ. Each set of smaller branches differentiates among

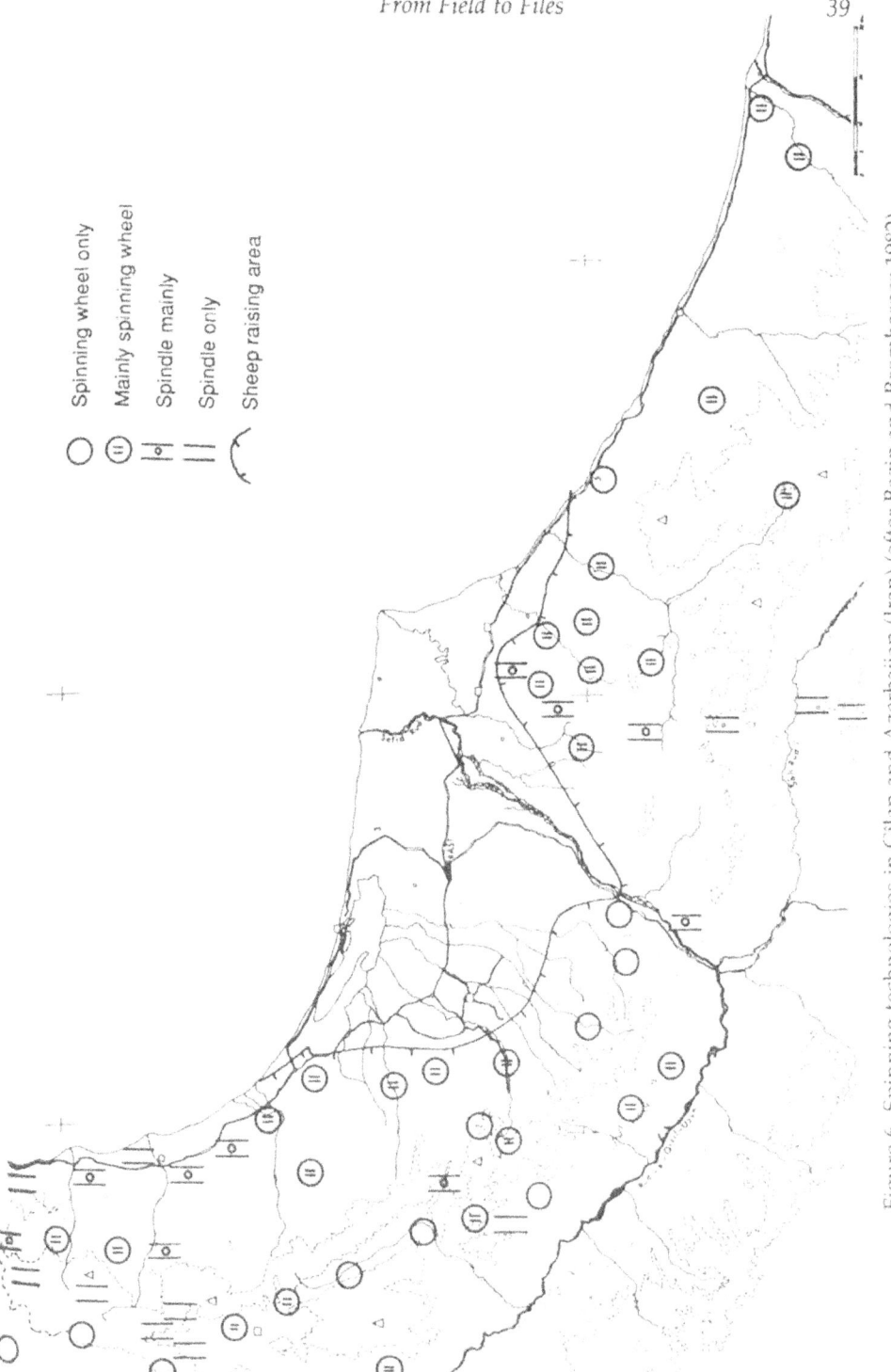

Figure 6. Spinning technologies in Gilan and Azerbaijan (Iran) (after Bazin and Bromberger 1982).

Figure 7. Types of houses in Tuscany (Italy) (after Biasutti 1952).

phenomena which were otherwise similar by criteria used in the classification of the bigger branches. Figure 11 shows a classification of airplanes by Quilici-Pacaud (1977) according to the number of their struts and control surfaces, and the position of the center of gravity. It also shows real historic filiation.

Figure 12 displays a classification, based on hierarchically ordered

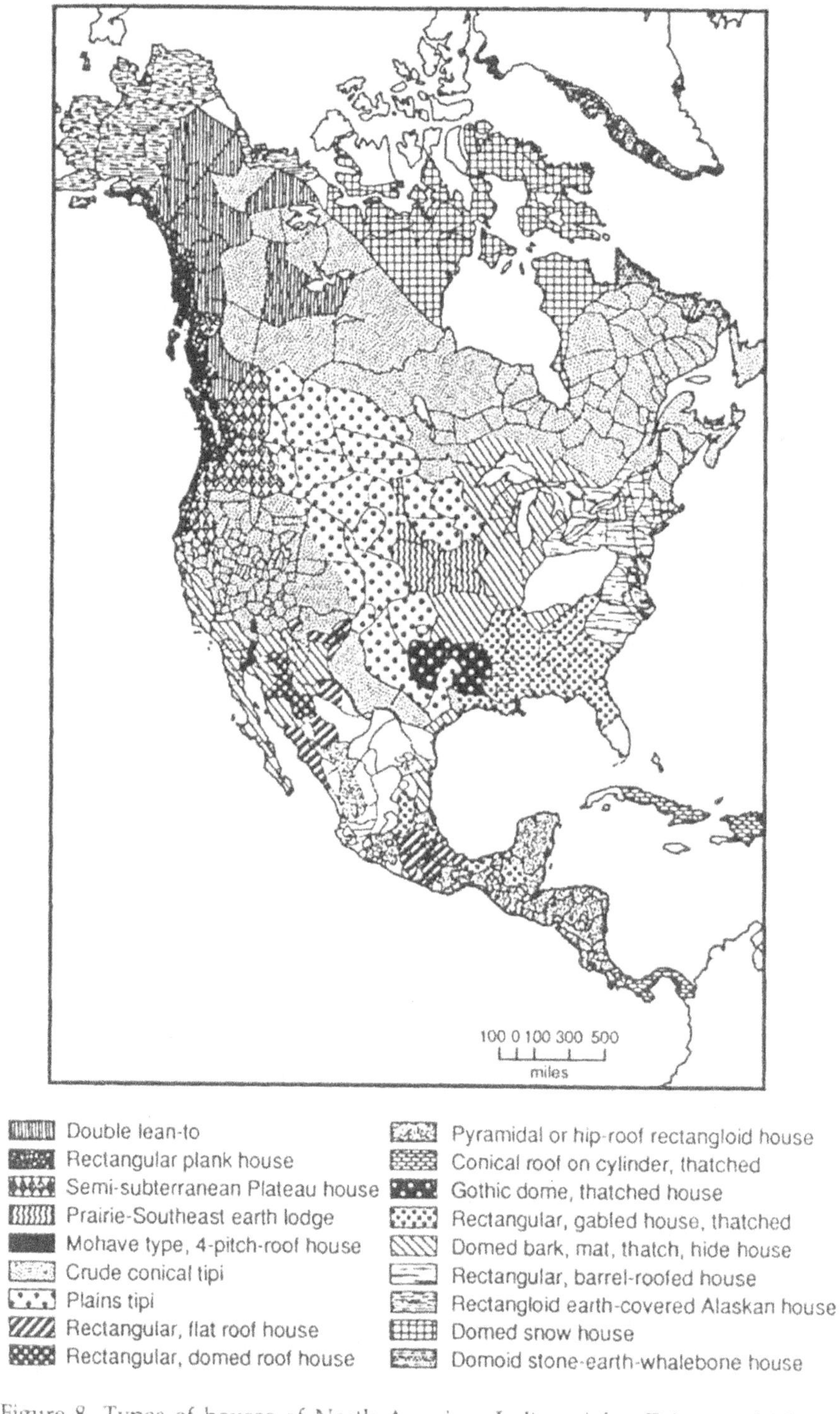

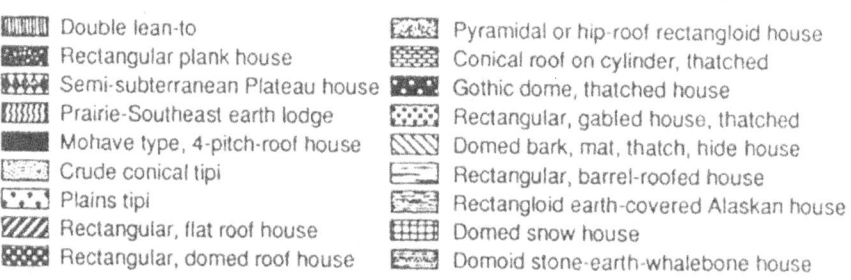

Figure 8. Types of houses of North American Indians (after Driver and Massey 1957).

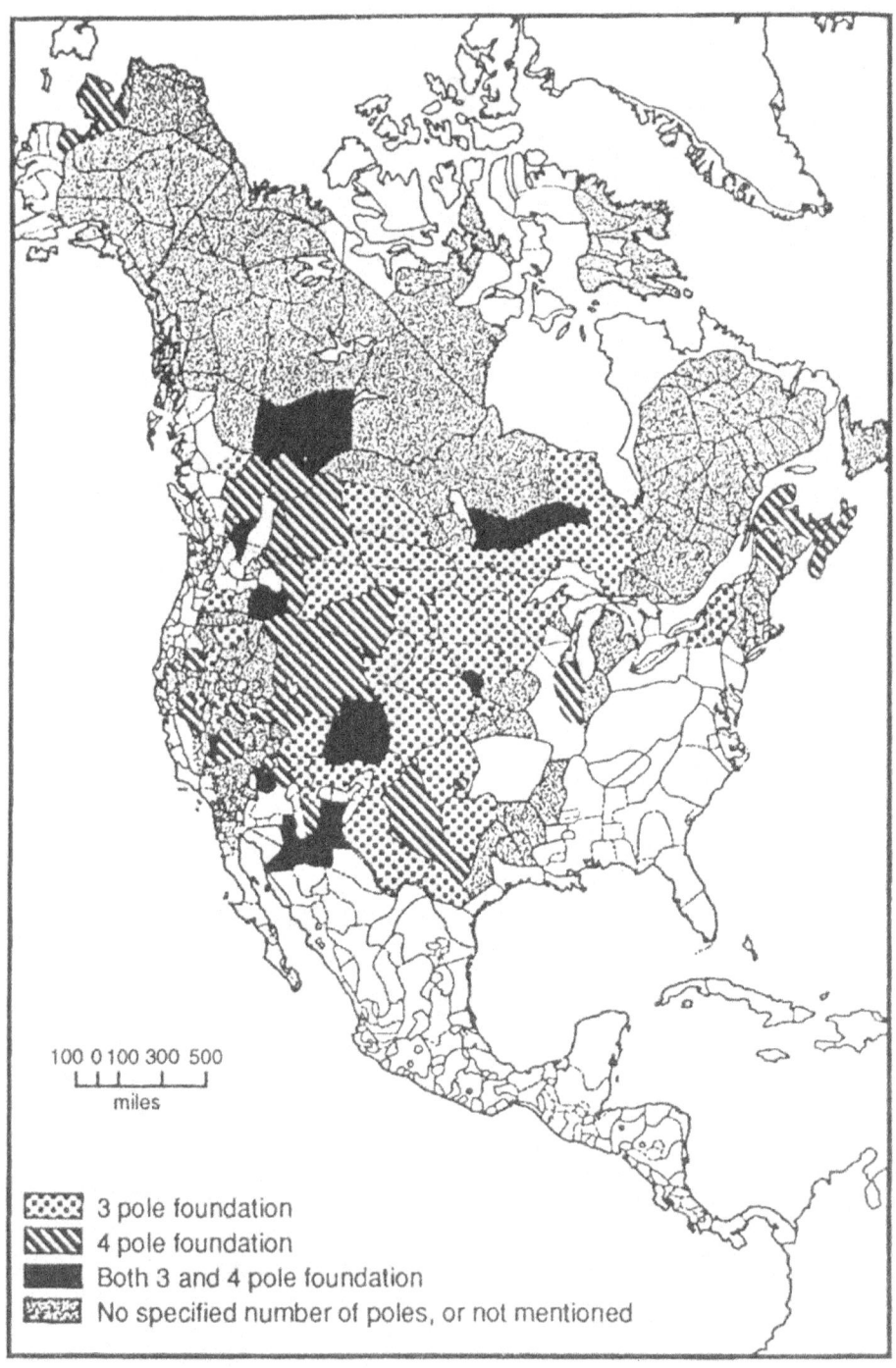

Figure 9. Distribution of conical and subconical houses of North American Indians (after Driver and Massey 1957).

Figure 10. Dominant house-building division of labor among North American Indians (after Driver and Massey 1957).

TABLE 2
Frequency and Ways of Stirring Brine in 13 Salt Marshes on the Atlantic Coast of France

REGION	stir without fine-salt	gather the salt	break-up fine salt and stir	break-up fine salt without stirring	stir thoroughly every day twice	stir thoroughly every day once	stir thoroughly every other day collection day	stir thoroughly every other day non-collection day
Carnac	X						X	
Mesquer	X	X				X		
Guérande	X	X				X		
Noirmoutier	X	?	X	X	(X)	X		
Bourgneuf			X				X	
Bouin		X	X			X		(X)
Beauvoir			X				X	
St. Gilles	X		X			X		
Olonne		X	X			X		(X)
Talmont	X		X	X		X		
Ré	X	X	X	X	X			
Seudre	X	X	X		X			
Oléron	X		X		X			

Source: Lemonnier 1980.

criteria, of salt-making processes in New Guinea. This figure emphasizes the successive transformations of salty materials following the simplified operational sequences in which a salty product is extracted from each kind of material. Other tree diagrams could have been constructed to represent those particular technological processes that focus more on the alternative operations possible at each step (consumption on the spot, use as a cooking liquid, impregnation, combustion, filtration, evaporation, and so forth), or that instead focus on the variety of salt-producing materials available. The order is a result of choices made by the investigator. In the present case, the choice was to illustrate the different combinations of possible transformations found in New Guinea. This was not an arbitrary choice, as a salient feature of salt-making processes worldwide is that they basically involve the same set of operations, the *order* of which, however, varies widely. I wanted to show that many of the combinations illustrated by each "path" of the tree diagram are highly localized, which means that, except in a few cases, processes that could have been used were not.[3] This, in turn, illustrates technological choices.

This point leads us to make another observation. The particular

TABLE 3
Technological Distribution of Principal Types of Agriculture throughout the World

Uses of Energy for Agriculture	No iron tools	Agricultural Use of Metal and Machines		
		Iron tools	Vehicles	Machines
Human Energy	(1) Non-metal cultivation: Indian America, Oceania, (European neolithic)	(2) Hoe cultivation: Black Africa, some areas of Far East		
Animal Energy excluding horse	(3) Swing-plow cultivation in prehistoric Eurasia: recent Stone Age, Bronze Age, early Iron Age	(4) Swing-plow cultivation: Mediterranean countries, western and central Asia	(5) Far East: India, China Southeast Asia, etc.	
Animal Energy including horse			(6) Ancient Europe: (from late Iron Age - La Tène to 18TH C.)	(7) Europe and North America: 19TH and beginning of 20TH C.
Fossil Energy (combustion engines)				(8) Industrial farming

N.B. Systems of agriculture shown in parentheses are now extinct.

Source: Sigaut 1984:361.

classification used to describe New Guinea salt-making, like all other classifications, results from reference to a wider body of knowledge, namely, that of salt-producing processes all over the world. I was acquainted with the combinative nature of salt-production, and I knew that some kinds of processes were not found in New Guinea, such as the natural freezing of sea water (seen in Siberia) or the pumping of artificial brines made by dissolving underground rock salt (which requires extensive mechanization and knowledge of chemistry). Nor does the figure show some salt-making processes that could exist in New Guinea but to my knowledge are not found there, such as artificial production of efflorescences, direct use of salty grounds, evaporation ponds. This observation is also relevant to the making of tables and maps. The kinds of hoe described by Sigaut, the types of percussion defined by Leroi-Gourhan, or the features of houses mapped by Driver and Massey are all the result of both a hierarchical ordering of criteria and more general knowledge about the technique or process of concern.

Thus, the making of classifications is a dialectical process: each

TABLE 4
Classification of Dani Arrows

Functional	Linguistic		Physical
Fighting	Mate	AIK DOK	short barbs
		HUBUK BALEK	?
		ILJAKAPI	long barbs
		WULEPUGU	2 rows barbs
		MAGUJA BALEK	3 rows barbs
		APELE BILIK	3 rows barbs
		ATOKWEJAK	3 rows barbs
		OAK HOBO	tips of barbs squared
		BIAL ELABO	shaft long, square cross-section
		LISANI GUGUK	barbs little more than bumps
		ILIGI USAKAIK	two-direction
		GENEK	barbless
Pig Killing		WIM	bamboo; unbarbed; single point
Hunting	only birds	DOPO	knobby, root end
	only mammals	DAKHU	barbed; single point; heavier than any fighting arrows
	birds and mammals	MUGUI	barbed; 3-pronged
		DUAP	barbless; multi-pronged
		PHIDE GOBAK	reed; barbless;
		DALONA { SINILAK / PUTUK }	{ barbed; / barbless; } bamboo; multipronged

(Arrows on the Fighting group indicate "hardwood barbed" with "more precise distinctions not ascertained"; on the Hunting group indicate "hardwood".)

Source: Heider 1970:284.

tentative classification allows a better ordering of more general phenomena, which, in turn, points to better criteria. At least this characterization reasonably describes the actual development of the study of technology. The best classification is the one that best fits with a given anthropological question, such as "how does the technological process I am studying differ from, or resemble other technological

TABLE 5
Types of Percussion

Percussions		linear		punctiform	diffuse
		longitudinal	transversal		
Perpendicular	resting	1	2	3	4
	dynamic	5	6	7	8
	resting with striker	9	10	11	12
Oblique	resting	13	14	15	16
	dynamic	17	18	19	
	resting with striker	20	21	22	

Source: Leroi-Gourhan 1943:56–57.

processes, in the same society as well as in others?" Each branch of a tree diagram can always be expressed using a more detailed classification using other sets of criteria. In Chapter 4, I will address the assumption that the further one moves away from the trunk toward smaller branches, the more stylistic the features become.

One last remark refers to the common categories used in classification. As already noted, "hoe" or "swing-plough" are implicit categories in Sigaut's table, as are "plains tipi" and "crude conical tipi" in Driver and Massey's, or "filtration," "soaking," and "evaporation" in mine. Insofar as everyone agrees on a common definition of given materials, operations, products, or the like, everything goes well. But

48   *Elements for an Anthropology of Technology*

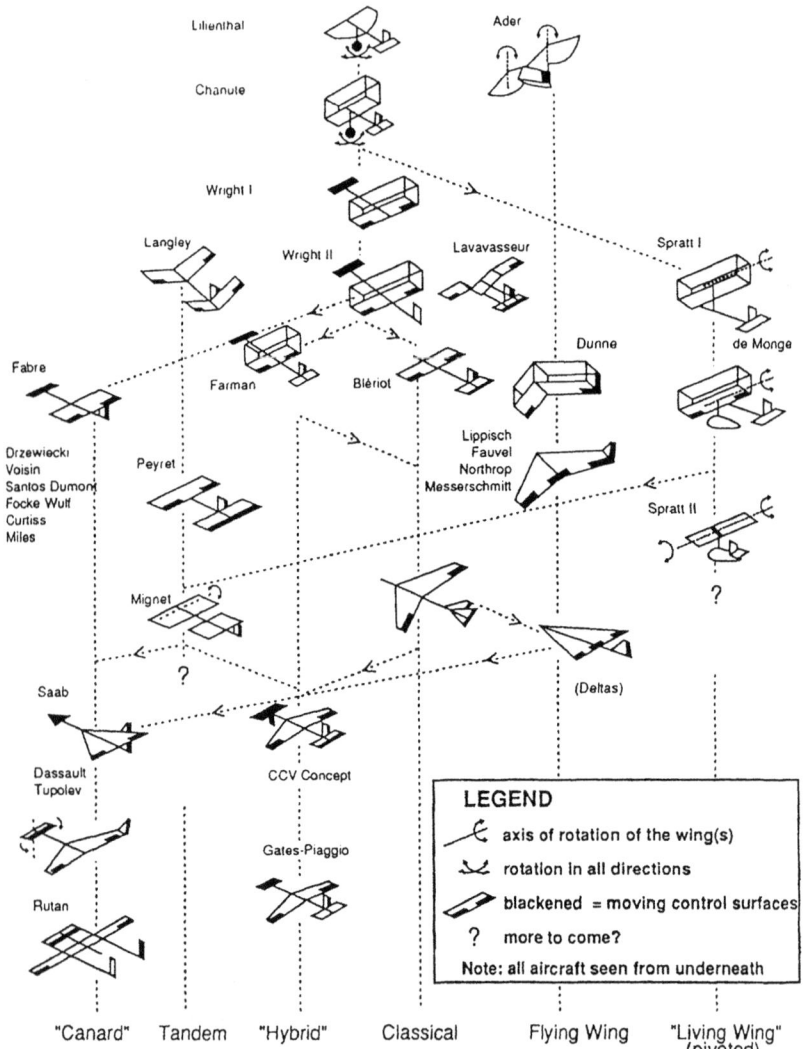

Figure 11. Classification of planes according to number of struts, control surface, and center of gravity (after Quilici-Pacaud 1977).

if a classification is expanded to include more precise operations, as it ultimately should, difficulties begin to arise. If the focus is on the similarities or differences between "Plains tipi" and "crude conical tipi," Driver and Massey's definition (a drawing) will not be sufficient (see Fig. 8 above). Definitions used in human engineering will not

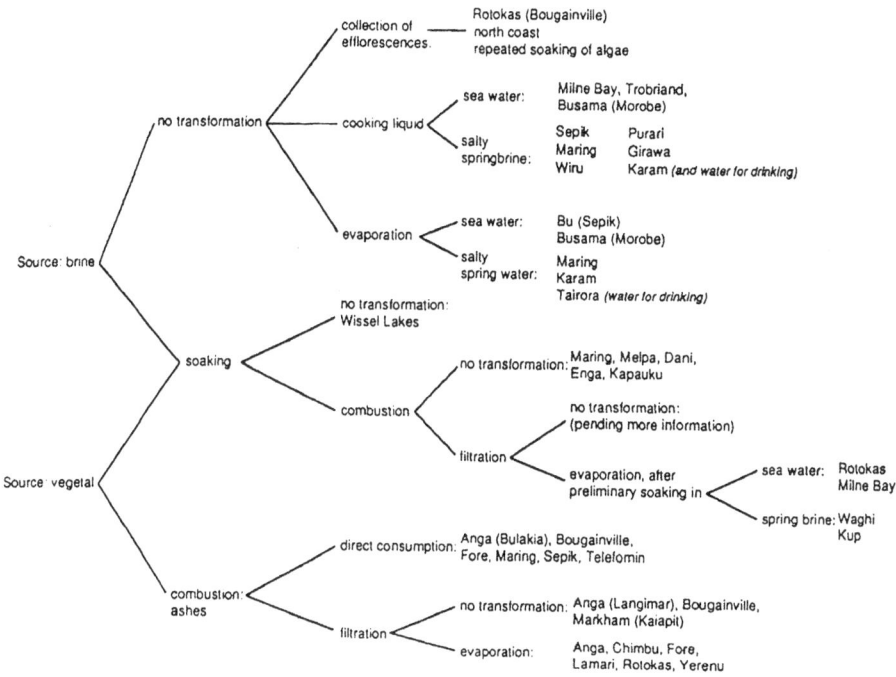

Figure 12. Steps in salt production in New Guinea (after Lemonnier 1984a:120).

fit, either. For example, the definition of the *therblig* (an anagram after the inventor's name—Gilbreth—used to refer to a standard operation) to "grasp" is: "begins when hand or body member touches an object, ends when control is gained" (Chapanis 1965:64). An expression such as "ends when control is gained" would not be accurate enough in a comparative study of "grasping," because there are many ways to "touch" an object, as there are many ways to "gain control." It is only by constantly refining and clarifying the categories we are using that common, unequivocal vocabularies and descriptive and analytical categories can be constructed.

Very few real classifications of technological actions can be found in the literature. More often, they are implicit, and thus quite hard to evaluate or use. The reason for this lies in the general underdevelopment of the ethnology and history of technological systems. Such classifications are nevertheless obligatory, not only because they make it possible to know what we are talking about, but also because they provide the only way to ask the kinds of questions that are crucial for understanding the transformations of technological systems. Among the many possible questions that one can raise, probably the most central is how to identify those choices involved in these transformations that are arbitrary. But first, we still need to demonstrate that such arbitrariness actually occurs in technology.

Chapter 3

Arbitrariness in Technologies

The aim of an anthropology of technology should not be the mere description of operational sequences and classification of technological features. These are only two means that must be developed in order to ask anthropological questions of technological systems. As already noted, the existence and scope of technological choices is an issue that goes far beyond the necessities embedded in the material world or in the internal physical logic of a given material culture. These choices arise where a technological issue confronts non-technological social phenomena. Such choices are arbitrary, at least from a technological-physical point of view. This chapter and the next are devoted to the search for evidence of arbitrary technological choices in two very different societies—the Anga of Papua New Guinea and our own.

SOCIAL REPRESENTATIONS OF TECHNOLOGIES AMONG THE ANGA

As is the case in other examples of material culture, the Anga technological system shows features that Leroi-Gourhan would have called *derniers degrés du fait*; that is, features whose functional character is negligible and which operate only in the symbolic dimension. The material and form of a skirt, for example, will express the gender, age class, initiatory stage, and ethnic group of the skirt wearer (Lemonnier 1984b). No less classically, some Anga artifacts, like arrows for instance, include features of limited functional significance, but whose distribution follows tribal boundaries (Lemonnier 1987). For example, for the simplest arrow type, with unbarbed point and round cross-section, only certain groups of Anga encircle the shafts with one or more bands of small, shallow, semi-lunate–shaped notches. The makers believe these notches will cause the shaft to break on impact, causing the point to remain in the wound, thereby increasing injury and mortality (Fig. 13). However, although their decorative function is clear, the notches are so small and shallow that their efficacy in enhancing breakage is doubtful.

But Anga techniques also display less superficial variations; that is, variations that are more closely tied to physical actions on material. The following pages will describe such variations as seen not only in a single type of object, but also for entire technical processes. Particular emphasis will be placed on the potential choices open to those who apply these techniques. Figure 14 summarizes the progression followed in this discussion.

Thus, certain "secondary" features which vary from group to group do have material functions—functions of a physical order—that are not negligible. This is the case for barbs on Anga arrows, for example, which are unknown in several groups, while common in others (Fig. 15); for Anga hearths, which are sometimes suspended, sometimes funnel-shaped (Fig. 16); or for the ligatures used in building houses, which in some groups have the appearance of an endless spiral while in other groups they are made knot by knot. It is important to note that in all of the examples cited thus far the solutions not used by a particular group are nonetheless well known to its members, because all of them have had ample opportunity to observe these solutions among their neighbors during trading or war expeditions. It is for this reason that I speak of "choices." A trait absent in one society is not necessarily unknown to it, and we may therefore conclude that this society has chosen not to utilize or produce it.

It also happens that we observe the absence (or the presence) of entire technical processes, and not simply specific traits of detail, in certain groups and not in others. Some of these techniques are genuinely unknown where they are not used. This is the case, for example, with certain techniques for hunting eagles or cassowaries, which, when described by the ethnographer to the incredulous representatives of most Anga groups, are taken as a good joke, provoking

Arbitrariness in Technologies

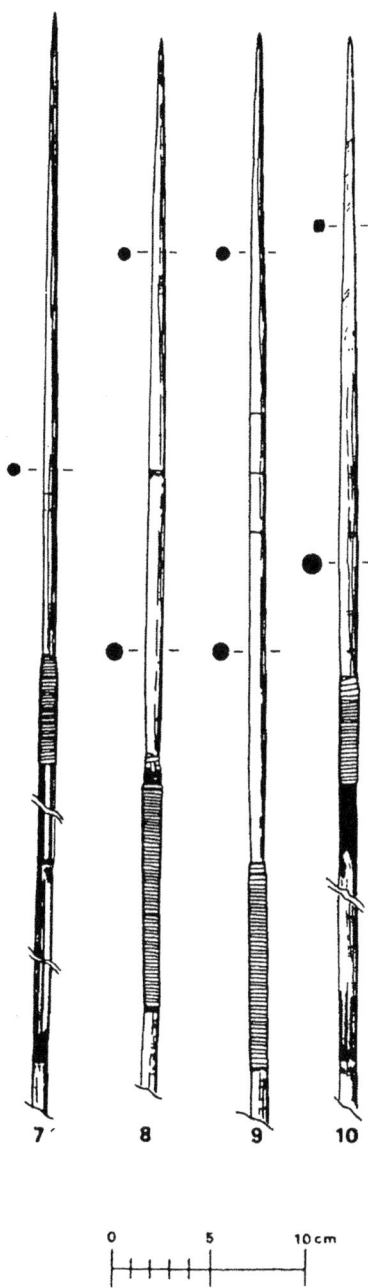

Figure 13. Arrows with grooves to allow breakage used by the Baruya Anga, Papua New Guinea (after Lemonnier 1987; drawing by Ph. Gouin).

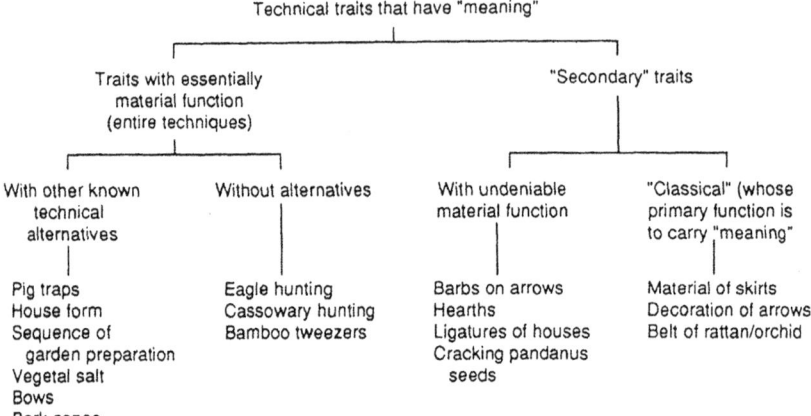

Figure 14. Nature and variability of Anga technical traits (after Lemonnier 1986).

gales of laughter, while these same techniques are common and well attested in other groups only a three-days walk away. Another example of this type is the use or ignorance of bamboo tweezers to turn over tubers cooking on the hearth. And a third example: during the preparation of "Polynesian ovens" to steam meat or tubers, heated stones must be moved to create a level bed in the oven or to cover the food as it cooks. Yet, while the Anga generally use sticks to extract the heated stones from the glowing coals, groups differ in how the stones are transferred to form the top layer of the oven. Some groups move the rocks barehanded, while others do it with the help of a carrying net to protect their hands. Informants belonging to groups unaware of the latter technique generally deny that it is possible to grasp a hot stone this way without burning and destroying the net.

Perhaps the most interesting technical traits and processes are those which, though known in a particular Anga group, are not used there. They pose in the clearest manner the problem of technical choices. In fact, the non-utilization of a given technical trait, as in the three examples which follow, cannot be explained by a failure of diffusion or of local invention. It is not for lack of knowledge, but indeed by choice, that a trait, or entire technical process, does not appear in the technical system under examination, since it is familiar to those who make no use of it and sometimes even described and commented upon by them. This type of rejection corresponds to

Arbitrariness in Technologies 55

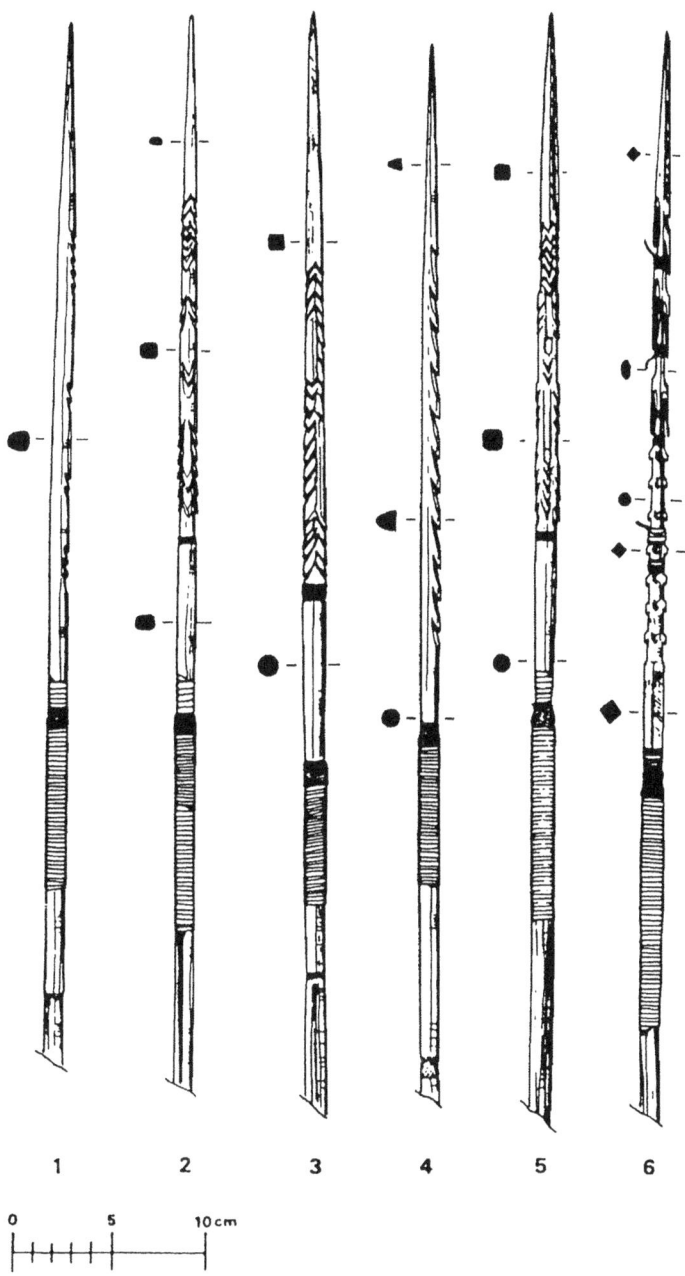

Figure 15. Barbed arrows used by the Baruya Anga, Papua New Guinea (after Lemonnier 1987; drawing by Ph. Gouin)

56 *Elements for an Anthropology of Technology*

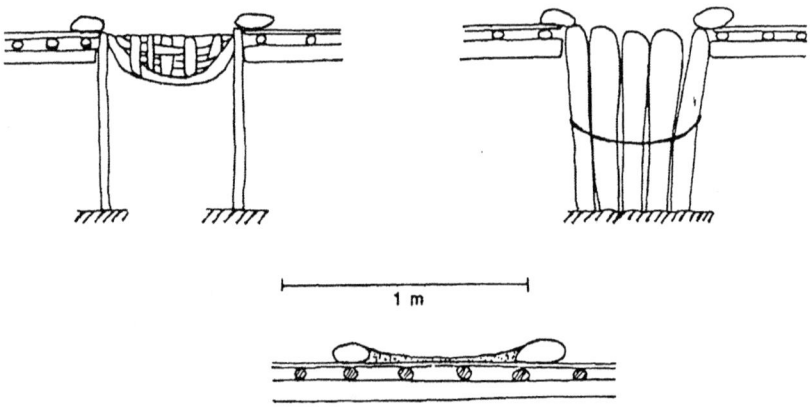

Figure 16. Types of Anga hearths (suspended, funnel-shaped, flat) placed directly on floor (after Lemonnier 1986).

what Leroi-Gourhan (1945:424–27) called an "unfavorable technical milieu," advancing as its explanation the incapacity of the technological system or the absence of material need to adapt a newly available trait. However, as we shall see using Anga examples, these technical choices clearly involve a purely social dimension, and any attempt to explain them without reference to these social aspects stretches the credibility of strictly material arguments to their limits.

Besides hunting with bows and arrows, the Anga make use of three types of traps to kill wild pigs. The first consists of a trench dug in the earth, the bottom bristling with sharpened stakes and the opening covered with branches (Fig. 17). The second is also passive and designed to transfix the animal by ensuring that it falls on a row of sharp stakes. This time the stakes are placed in a garden, set at ground level pointing towards an opening in the barrier which normally blocks access to the trap. Scenting a windfall, the pig leaps the barrier only to impale itself (Fig. 18). The third is a dead-fall trap, consisting of two or three strong cudgels which strike together in a passage where the animal is first immobilized, crushing its head and body with their weight (Fig. 19).

In the majority of Anga groups, these traps are all used, the substitution of one for another depending variously on the nature of the terrain and even more on the inspiration of the trapper. The Langimar are an exception. Members of this group can name without difficulty the ten pieces that make up the dead-fall trap, they can describe its functioning, and they can even make a rough sketch; but

Arbitrariness in Technologies 57

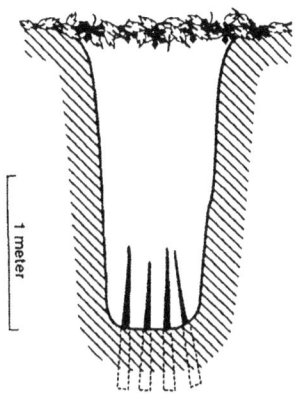

Figure 17. Pit-trap for pigs used by the Anga, Papua New Guinea (after Lemonnier 1986).

they do not use the device. To the classic "our ancestors did not use it" response—the polite answer to any stupid question from an ethnographer concerning the origin of things—they nonetheless add that the Kapau, whose closest villages are only a two-or three-hour walk away, or the Menye, whose houses are easily seen across the river, currently use this trap. I shall return to this association of material culture with ethnic identity later. Suffice it to note for the moment that a perfectly understood technology is voluntarily ignored by members of a given group.

Another example of choice is illustrated by the distribution of arrow types among the Anga. I have already indicated that, on first analysis, the presence of barbs may appear to be a detail of style. Figure 20 shows the distribution of barbed and unbarbed arrows without reference to subtypes. We see clearly that the northern Angans use both barbed and unbarbed arrows, and the six groups occupying the southern Anga area do not use barbed arrows. There can be no doubt that at least the Kokwaye, Menye, and Kapau, who have physical contact with the northern groups, have had many occasions to note the superiority of the barbed arrows shot at them by their enemies. Nevertheless, they have neither fabricated nor imported them, proof that a reason other than "technique" here opposes the adoption of a device more deadly than the ones they habitually use. I should add that no correlation exists between the use of barbed arrows, the types of wood available, or the game hunted (Lemonnier 1987). The existence of a choice in this case seems indisputable.

58   *Elements for an Anthropology of Technology*

Figure 18. Pit-trap constructed at opening in fence, Papua New Guinea (after Lemonnier 1986).

The case of Anga houses exemplifies yet another quasi-arbitrary choice of a particular technical solution by one or several groups. Among the dozens of criteria one could use to characterize the typical house type of each group, the number of walls making up the enclosure immediately draws one's attention. In the northern groups (Simbari, Baruya, Watchakes, Kokwaye, Menye, Yoyue, Kawatcha, Langimar), the houses take the so-called "hive" form, their thatched roof descending to the level of the floor over the pilework or, indeed, lower (Fig. 21). They have a circular wall made of double, vertical layers of bark or Pandanus leaves and a single door-opening.

Among the Ankave, Ivori, Lohiki, and Kapau, the houses are lower and larger; and, above all, most of them have two enclosures: the exterior one delimits a passage whose interior wall encloses the only part that is lived in, most often circular. The floor of this room is

Arbitrariness in Technologies 59

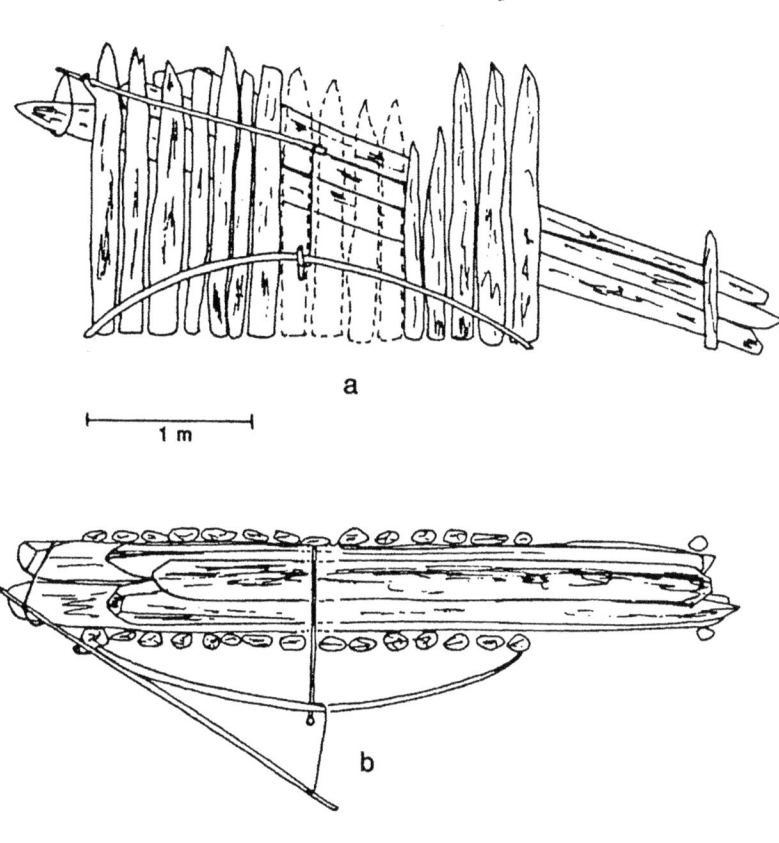

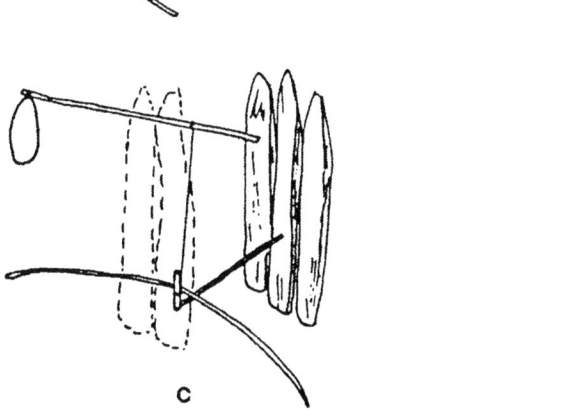

Figure 19. Anga dead-fall trap for pigs (after Lemonnier 1986). *a*, side view; *b*, top view; *c*, detail of trigger.

60   *Elements for an Anthropology of Technology*

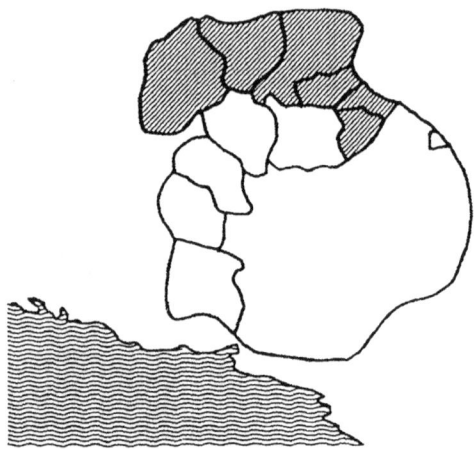

Figure 20. Distribution of barbed arrows (shaded area) among the Anga (after Lemonnier 1987).

Figure 21. Baruya Anga house with single enclosure wall (after Lemonnier 1986).

elevated, and reached by a door that generally is situated in the first two meters of the passage, to one side or the other of the external opening (Fig. 22).

When I first encountered a house of this type at Kanabea, in Kamea country, a high altitude (1,600 m) area known for its long periods of bad weather, humidity, and low temperatures, I thought that the presence of the double wall was quite logical, as it afforded protection from the cold and retained the heat from the hearth. My logic, however, did not take into account the fact that other groups (Baruya, Watchakes, Yoyue) build houses with only one enclosure, and yet it is not uncommon to see the thermometer in these regions at less than 10°C at dawn. Moreover, houses with double enclosures are also found at elevations around 400 m on the last slopes bordering the Gulf of Papua, where the nights are much warmer. In this case, the interior compartment has a skylight. It is only in the Papuan lowlands that groups like the Ivori, Lohiki, and Kapau build houses with no wall other than a half partition. The presence of houses with single enclosures in some places and double walls elsewhere is therefore no proof of environmental determinism. At equivalent temperatures or humidities we find, depending on the group, one type or the other. Yet the Kapau, for instance, know perfectly well that their northern neighbors build houses with one enclosure; reciprocally, the Kokwaye, Menye, or Langimar have had occasion to see, at one time or another, double-enclosed houses. Only the Ankave traditionally built both types of houses. In the absence of any clear correlation with environmental factors, I speak once again of choice.

Whether dealing with relatively secondary traits or, on the other hand, with entire branches of technical activity, the collection of examples cited so far points toward an important initial result. All else being equal, certain modes of action on material, which I call technical traits, are differentially distributed among the twelve Anga groups considered here, and their distribution affirms that these technical variants are not merely responses to different material environments; instead, they appear to be arbitrary choices proper to some groups but not to others. What is remarkable about this result is as much the nature of the traits considered as the range of their variation. Their principal function involves a physical action on material rather than the transmission of information, and sometimes involves entire techniques. To me, the intensity of their variations in an apparently very homogenous cultural context is amazing. A final example will illustrate the surprising diversity that results from the same type of action on material in neighboring societies.

In seven highland Anga societies cultivating the same vegetable

Figure 22. Anga house with double enclosure wall (after Lemonnier 1986).

species in largely comparable ecosystems, the last phases of garden preparation bring into play an operational sequence whose final three stages are: (1) the burning of plants cut down during land clearing, (2) the construction of a barrier intended to protect the garden from the incursion of semi-domestic or wild pigs, and (3) the planting of seeds or cuttings (Lemonnier 1982). Yet, these three important operations follow a variable order depending on the society considered; three configurations can be observed:

1) burning-barrier-planting (Baruya, Watchakes)
2) burning-planting-barrier (Langimar, Simbari, Yoyue)
3) barrier-burning-planting (Menye, Kapau)

It is quite surprising that tasks so specific, and so closely associated in the same highly specialized gardening operations, may constitute sequences so variable in societies sometimes less than a day's walk apart and exploiting similar ecosystems. Nonetheless, such is the reality of Anga agriculture. One can easily imagine, therefore, the latitude left to technical creativity in these societies for operations which *a priori* are less tightly constrained than agricultural tasks.

Figure 14 above illustrates the structure of the different examples discussed. In it, we find the normal continuum of *degrés du fait* or, if one prefers, a progressive transition from technical traits whose principal function is to transmit information (*derniers degrés du fait*) to others whose primary function is to act on material. In the lower branches of the tree diagram in Figure 14, I have opposed traits for which no other technical solution is known or available to those which correspond to incontestable choices made by the Anga. This opposition does not appear in the upper branches of the tree (classic "secondary traits"), which are all traits that each Anga group or ensemble of groups know to be their own.

But study of the Anga technical system reveals not only operations that are more arbitrary than we might have expected; it also brings to light the existence of an organizing principle behind the diversity and arbitrariness of these technical choices. Having noted the differential distribution of several technical traits, there is occasion to wonder if this occurs by chance. The map of the distribution of houses having one or two enclosures provides the first element of an answer to this question (Fig. 23), for we see here two clearly delimited zones within which the trait considered maintains the same form. In the northern part of Anga territory, the houses possess only one circular wall, while in the south they have a passage and double door. Hence, we have a continuity in the distribution of each trait that allows us to suppose that it does not occur by chance but, on the contrary, reflects the same "point of view" shared by several groups for the technique considered. But there is more. If we compare the distribution of house types among Anga groups, as shown in Figure 23, with the distribution of barbed arrows (see Fig. 20 above), and then with the distribution of bow types (Fig. 24), we observe that zones which are opposed for a given trait are in very large measure also opposed for the other two.

Such parallel variations would not be at all surprising if the technical traits considered were functionally linked. It is logical, for example, that the form of a cake of vegetal salt depends directly on the mold in which it is crystallized, or that the types of wound inflicted on game are linked to the nature of the traps or weapons used. But there is nothing of the sort in the present case; the two Anga bow types can shoot any type of arrow with the same effectiveness, and whether they are barbed or not has no relation to the cross-section of the bow that shoots them. We would also have trouble finding common technical antecedents to the form of both the houses and the bows. Finally, the probability that three traits, capable of taking two different values, taken at random, would have exactly the same dis-

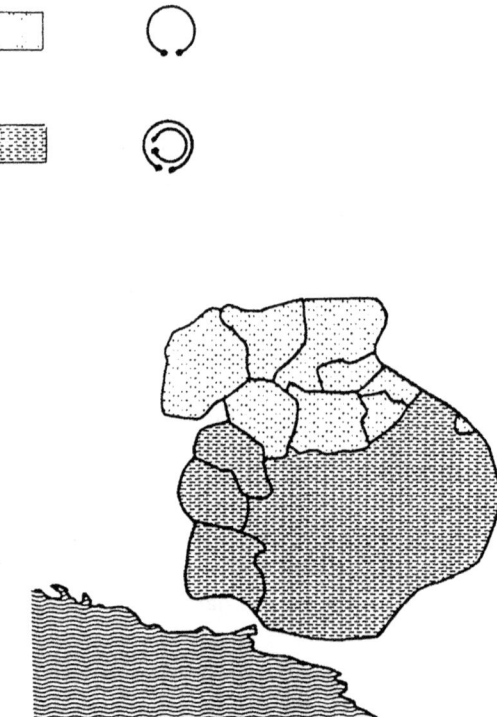

Figure 23. Distribution of Anga houses with single and double enclosure walls (after Lemonnier 1986).

tribution in twelve groups is very low, so low that it is perfectly reasonable to admit that chance has nothing to do with it.

Having eliminated both the hypothesis of a functional link and that of random distribution, we return to the hypothesis of an ordering and classification of the technical domain expressed by choices, ultimately unrelated to what the natural environment or a strictly technical (material) logic would lead us to expect. The simultaneous presence in one or several groups of a series of techniques not functionally linked would then "simply" reflect the application, most often unconscious, of one or several indigenous classifying principles.

A rather different and obvious first explanation would be to suggest that the patterns one observes today in Anga material culture are the result of historical processes. In other words, functional reasons might have led to the adoption or rejection of certain technological

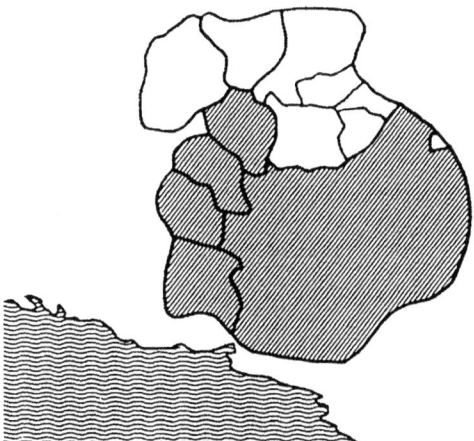

Figure 24. Distribution of Anga bow types: shaded area represents non-oval (semi-lunar and subtriangular) (after Lemonnier 1987).

features at a time when some Anga tribes were living in slightly different kinds of ecosystems, or had different neighbors. The functional-physical logic of these adoptions or rejections might no longer be visible now, after centuries of migrations, contacts, or internal evolution of their technological system.

Two comments should be made here. First, as the recorded history of the Anga extends back only to between 1930 and 1955 (depending on the tribe), any historical understanding of what happened in the past must be coupled with archaeological investigations that have not yet been undertaken. Moreover, given the climatic conditions and resultant poor preservation of material culture in the New Guinea Highlands, archaeology may not be able to contribute much of the needed information. Furthermore, oral tradition emphasizes primarily the common origin of all Anga groups (probably several thousand years ago, according to glottochronology and genetic studies).

Second, whatever the functional background of the situation may have been, we are talking about the present patterns of making and using the artifacts, and performing the operational sequences. These patterns still result in a striking non-random and non-physically explanable distribution of objects and ways to make and use them. It must be concluded, then, that there are still representations of technological phenomena underlying these technological differences or variations, and the physical consequence of this is far from being neutral, as I have shown.

The crucial question, then, is to what extent social representations affect the development and performance of technological action. And, more generally, it is crucial to know up to what point arbitrary choices influence the transformation of technological systems. Jumping from New Guinea to the high-tech industry of our own society, we shall now look at whether such strange phenomena are limited to so-called primitive societies.

A GLANCE AT AIRPLANES: ARBITRARY CHOICES IN SO-CALLED HIGH-TECH SOCIETY[4]

The following example deals with a particular technology of modern industrial societies. It shows that representations of technologies play a part even in our high-tech world, and explores the extent to which representations of technology play a part in decision making. This example concerns the history of aviation, a domain in which one would think the constraints of the natural world weighed particularly heavily, and a domain which has been handled by highly specialized and talented engineers from the beginning.[5] A plane would not seem to be a technology with much latitude for technological choices, except for ones that had been previously planned, and only superficial decorations such as figures and letter for identification, airline logo, or upholstery would seem likely to carry stylistic information. As in the case of an adze or a bear-trap, the shape of a plane expresses several of its physical functions, particularly those adapted to flying and carrying a payload: wings for lift; fuselage for payload; engines for movement; ailerons, fin, rudder, and elevators to stabilize and direct the flight. Yet, if one looks at the history of aviation, and considers successful aircraft that flew and met their designers' expectations, the diversity of technological answers is striking.

Consider, for example, the location of engines on propeller-driven civilian transports, already a limited sample: the propeller can be tractive or propulsive; and the engine(s) can be located under the wing, inboard on the wing, above the wing, at the bow of the fuselage, in the fuselage and inboard on the wings, in the fuselage and under the wings, at the bow and at the back of the fuselage, in each of the wings of a multiplane, between the wings of a multiplane (Fig. 25), at the back of the fuselage (Fig. 26), inboard on the wings and included in the fin (Fig. 27). Engines may number from one to eight and even up to twelve.

A comparable diversity of technological answers can be found for other functions as well: for the wings can vary in shape, location, number; the horizontal stabilizers can occur or not occur, and at the back or at the front of the body; the landing gear can conform to various designs. Passengers also are sometimes found in quite surprising places: in the floats of a hydroplane (Fig. 28), in two fuselages (Fig. 29), and even in the wings, partially, or totally, if we include the once envisaged civilian version of a Northrop bomber (Fig. 30).

Thus, if one focuses on a single type of plane, for instance the high-winged, twin-engine light transport, clear differences appear among aircraft designs. The general case is an airplane with flat or in-line engines, tractive propellers, and body continuing back to the horizontal stabilizer. But, we also find propulsive propellers, radial engines, and short fuselages, not to mention the flying-boat versions or the use of turboprops.

If one narrows the scope of discussion, new fields of variability appear, a common feature of studies of technology. Nevertheless, all variations discussed so far deal with functional aspects of the airplane, and not with style or the communication of information between social groups. Every technological trait mentioned contributes fundamentally to the functioning of the aircraft, that is, to its ability to move a payload through the air. *A priori*, none of these traits marks any particular social or ethnic identity, nor do they carry any kind of information, except perhaps for a specialist as an indirect clue to the designer's particular school of thinking: planes with huge turboprops and contra-rotating blades are always Soviet, for instance. These variations do not even give clear information on a given age of aeronautics, since, with one or two exceptions (multiplanes, great number of engines, passengers located in floats), all of the foregoing designs are still commonly used today. All of these technological features nevertheless display great diversity, which does not (or did not) prevent these planes from flying normally. Indeed, some of them flew faster and farther, and with greater safety and less expense, than others, but all of them are (or have been) efficient technological artifacts coping with their users' requirements.

While the designers of each were no doubt convinced that they had

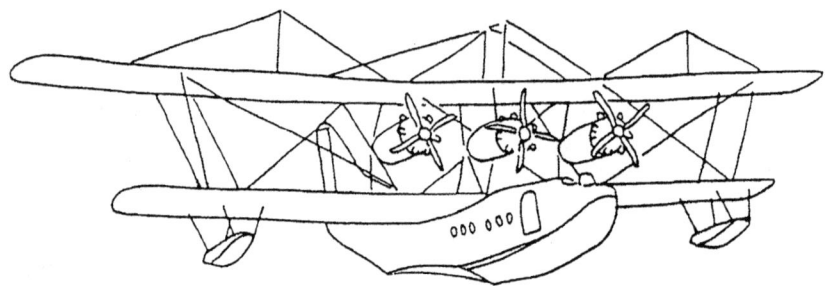

Figure 25. Engine between wings of Short Calcutta multiplane (drawing by Ph. Gouin).

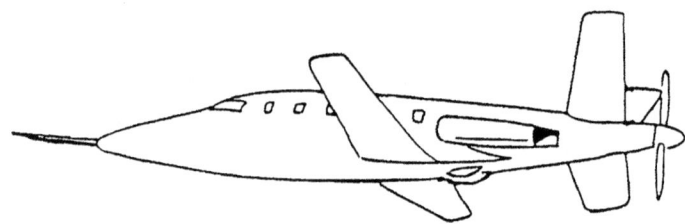

Figure 26. Engine at back of fuselage of Lear Fan 2100 (drawing by Ph. Gouin).

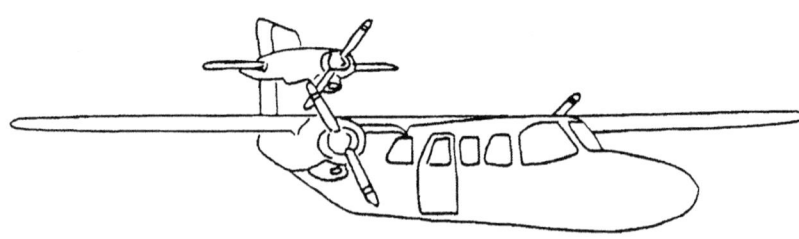

Figure 27. Engine inboard on wings and included in fin of British Norman Trislaner (drawing by Ph. Gouin).

Arbitrariness in Technologies 69

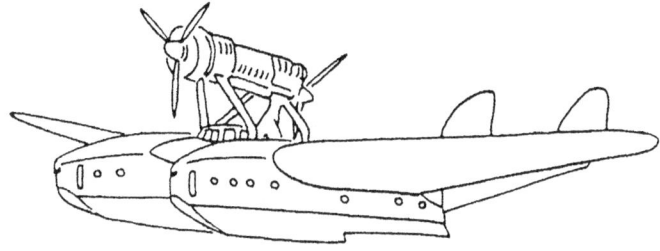

Figure 28. Passengers in floats of SIAI Marchetti S.M.55X hydroplane (drawing by Ph. Gouin).

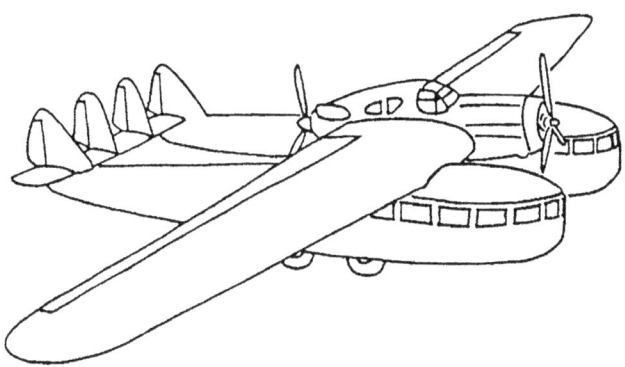

Figure 29. Passengers in two fuselages of Blériot 125 (drawing by Ph. Gouin).

found the best solution to various design problems, and while there have been obvious improvements as a result of the growth of technological know-how, nonetheless these changes only concern very specific domains, such as relative cuts in costs, as well as improvements in aerodynamic efficiency and weight of aircraft, power output and fuel consumption of engines, and of course safety. They obviously have led to gains in speed, range, and capacity, and to reductions in accidents and costs. However, these improvements hardly concern the general choices reviewed here. Whenever radical changes in the basic design of planes have been proposed, they usually have been rejected.

Thus, for example, although the "flying wing" was first designed

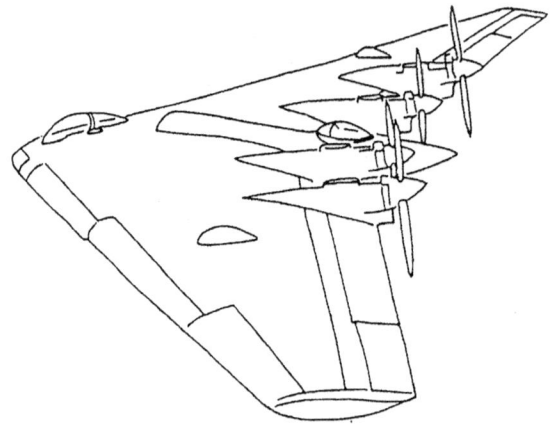

Figure 30. Passengers aboard a Northrop YB49 "flying wing" (drawing by Ph. Gouin).

in 1910 by Dunne, faced with more *conventional* opponents none has reached series production yet, even though Northrop's prototype dates back to 1928 (Taylor 1981:34–35; Angelucci and Matricardi 1983:396; Wooldridge 1983:49–52; Sweetman 1985).[6] The advantages of this particular design are obvious, however: the suppression of fuselage and horizontal stabilizers create an aircraft that is less likely to stall or go into a spin (Wooldridge 1983:54); and important savings in structural weight allow the plane either to carry the same payload farther, by having more fuel on board, or to increase the payload for a given distance.

Locating the horizontal stabilizer in front of the main supporting wing ("canard" foreplane) also diminishes the weight of an aircraft for comparable lift capacity (Fig. 31). Although they date from 1909, and even from the Wright brothers' Flyers gliders (Hooven 1978), canard foreplanes are still only being used on a few fighters, and on amateur do-it-yourself planes, or on prototypes. Although they form a limited case, one should note that "wingless" planes are even more rare (e.g., Lifting Body Martin Marietta X24).

One wonders therefore why particular technological features are held back by aeronautic designers. Are these formulas or devices dismissed because of their inefficiency or cost; or, have they instead been by-passed for non-physical reasons having to do with fashion or the ideas people have of what an airplane is "supposed" to look like? In other words, are they representations of this particular tech-

Arbitrariness in Technologies

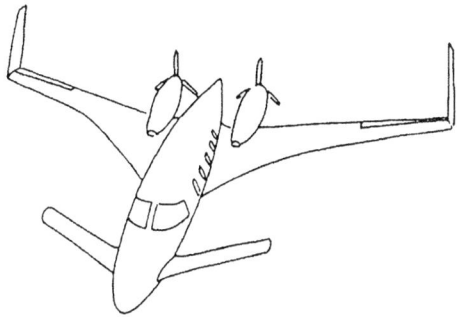

Figure 31. Canard foreplane (Beech Starship) (drawing by Ph. Gouin).

nology that have nothing to do with the physical or mechanical principles that aviation brings into play, nor with economic considerations?

Obviously, one cannot make any design fly. The laws of aerodynamics and the air-resistance of materials create constraints that designers and builders have to respect, using the know-how of the time. Thus, the Phillips multiplane (1904) was a complete failure (Fig. 32), even though in 1907 the same inventor was to make the first powered (and controlled) flight ever to take place in England (157 m) with a more complex version of his engine-driven "venetian blinds." Likewise, an apparent success (except that it flew in the opposite direction to what its classical appearance led one to expect), Dixon's Nipper (Fig. 33) was nonetheless a resounding failure (Taylor 1981:9, 11), despite the fact that Santos-Dumont had been the first man ever to fly more than 25 meters with his 14-*bis*, a biplane aircraft sharing with the Nipper the use of a canard foreplane and propulsive rear engine. The Phillips multiplane and the Nipper were just too close to the borderline of aeronautical laws.

On the other hand, there are some aircraft that had no success for reasons other than their ability to fly. I am not talking here of planes that were outclassed by their opponents, or whose original purpose was no longer of interest when they were ready for series production (a common occurrence for military aircraft). Rather, I am now considering aircraft that are as efficient as, and sometimes more efficient than, their opponents. Two planes provide an illustration: Cessna's C137 Skymaster (or "push-pull") (Fig. 34) and Mitsubishi's MU-2 (Fig. 35). Both were failures in spite of their unquestionable qualities and possibly even superiority over similar aircraft. For example, the

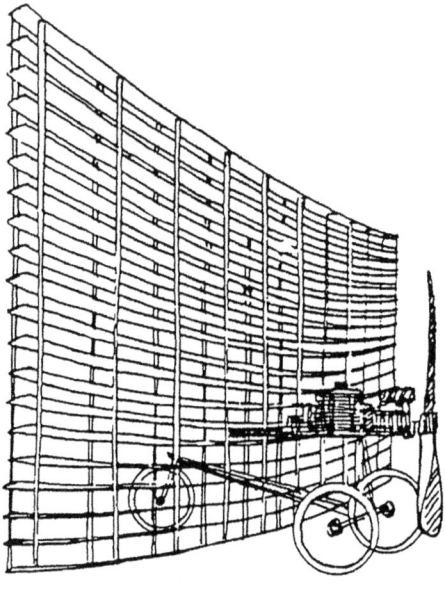

Figure 32. Phillips multiplane (drawing by Ph. Gouin).

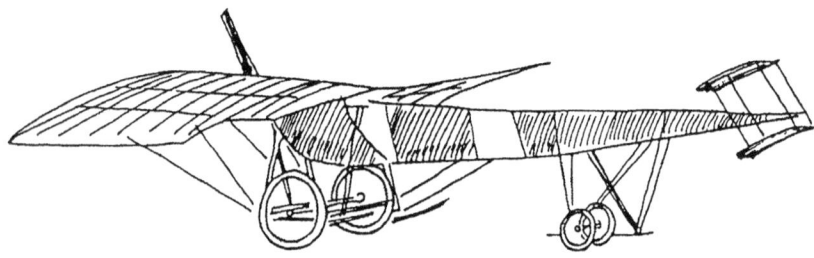

Figure 33. Dixon's Nipper: flies from left to right (drawing by Ph. Gouin).

longitudinal location of the Cessna's two engines maintains the plane's lateral stability in the event that one of its two engines should fail, whereas ordinary twin-engine planes have a tendency to spin, with obvious fatal consequences if this happens during take-off. Notwithstanding this basic quality, mistakes in the use of the aircraft (e.g., attempts to take off on only one engine), as well as its unusual appearance, and the kind of safety it afforded which did not fit most male pilots' rugged, no-nonsense image of how a twin-engine plane should perform, made the Cessna C137 a commercial failure. According to an observer, "one of the biggest problems in marketing the Skymaster was overcoming its image as an airplane for wimps. Centerline thrust was not macho. . . ." (Thurber 1985:40).

Mitsubishi's MU-2 was also an odd-looking machine. But once again, the aircraft was definitely a successful piece of technology, as shown by its ability to take off from short grass airstrips as well as its top speed which surpassed its best competitors by 100 knots (180 kph). On the other hand, its small wings (which allowed its high speed), and special use of high-lift devices (flaps), gave it an unusual silhouette and required new piloting techniques all of which led to its commercial failure: "Along the way the MU-2 became controversial because it is unique" (Mac McClellan 1985:40–42). These two contemporary examples show how the social representations of the aircraft itself, on the one hand, and the weight of routine in operating it (that is, the technological knowledge of the time), on the other hand, led to the dismissal of shapes which, although quite odd, were perfectly adapted to their purpose. In this case, skilled technician-users (the pilots) were the ones who turned these aircraft down. Moreover, no plane-maker today is likely to reintroduce these particular bright ideas.

Most designers produce only machines that fit their own representation of what an aircraft should look like. In his introduction to *Fantastic Flying Machines*, Taylor (1981:7) writes: "While a few prove dead-ends technically, others were inspired, only their radically *unusual* appearances preventing series production" (emphasis added). In the same way, D. Kücheman, the head of the aerodynamics department in the Royal Aircraft Establishment between 1966 and 1971, when the supersonic transport Concorde (SST) was being designed, noted the "traditional reluctance to accept new concepts on the basis of their technical merits" (quoted by Owen 1982:39).

The particular design that Kücheman had suggested violated a basic principle of aircraft design, a law proposed by G. Cailey (1773–1857) in the nineteenth century, according to which, the main functions of a plane have to be effectuated by separate components. Kücheman had proposed a partial integration of wings and fuselage, so that the latter would contribute to the total lift. Though not incorporated into the Concorde, this proposal was employed in the Rockwell B1B bomber, a plane designed for the greatest possible efficiency. The development of the SST and B1B bomber illustrates a more general observation: the higher the performance demands of an aircraft, the narrower the scope of variability in its design. We

74   *Elements for an Anthropology of Technology*

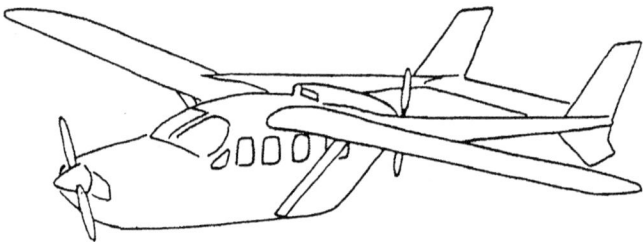

Figure 34. Cessna C137 Skymaster (drawing by Ph. Gouin).

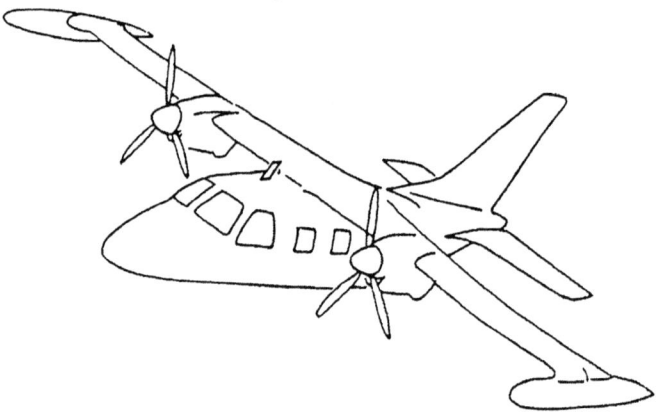

Figure 35. Mitsubishi MU-2 (drawing by Ph. Gouin).

have to keep in mind that in the beginning of the 1960s, when the SST project was taking shape, it was still a critical and dangerous exercise to reach supersonic speed, even with a military plane. At that time, the B58 Hustler, an American bomber, would disintegrate in mid-air whenever an engine failed at supersonic speed, even though it was non-pressurized to eliminate the risk of explosive depressurization. Seemingly insurmountable difficulties lurked for engineers who wished to fly a hundred passengers safely across the Atlantic at Mach 2 in a pressurized cabin. Facing new and extremely difficult problems—how to design a shape that would reduce the heat resulting from kinetic energy but still handle well at low speeds, or withstand tremendous differences in pressure between the interior and exterior of the cabin, or how to cool the cabin's interior and internal systems—the scope of possible designs narrowed considerably. Multiplanes or planes with fixed landing gears were no longer viable options. The "slender delta" plane became the choice to use at Mach 2, and technological choices then had to switch to the myriad problems to be solved to make a slender delta SST fly (Fig. 36). The laws of the physical world, as well as the technological knowledge of the time, left but a narrow margin within which to maneuver.

A comparison between Schneider Coupe racing hydroplanes (1913–1931) and the best fighters of World War II shows a similar phenomenon of convergence. The search for an optimum between power and speed led, given the time and available technological knowledge, to similar shapes and configurations (Fig. 37).

Thus, a look at modern advanced technology brings us back to a well-known principle in the anthropology of technology: the weight of "tendency" (*tendance*), in the words of Leroi-Gourhan (1943:27). To him, the more a technological feature is concerned with the *premiers degrés du fait* ("first degrees of fact")—those dealing with an action on matter or energy—the less it is subject to variation. When this technological action takes place among technological principles which are themselves very compelling, the scope of possible variations is narrowed ever further.

But within these limits imposed by the physical world, choices are still possible, the logic of which has nothing to do with the "objective" knowledge brought into play, nor with any style or communication of information. For instance, onboard computers and "fly-by-wire" technology have allowed us to fly aircraft that are far less stable than the flying-wing of the canard plane. At a time when a two- or three-percent savings in fuel may determine the success or failure of an aircraft among airlines, one or the other of these nontraditional formulas could perhaps lead to a significant increase in profits. They are not on anyone's agenda so far, however. Thus, in 1985, when Boeing presented the design for a 150-passenger canard plane among proposals for projects equipped with propfan (transonic propeller) engines, the specialized journals immediately suspected a tall tale rather than a workable project. Among the possible designs for a propfan plane, the canard has only a small chance of being developed, mainly because it does not look like a classical flying machine. Likewise, among the stealth bomber projects of the 1980s, the Northrop flying-wing proposal was said to have been put forward just to trick the Soviets.

The social representations playing a role in these choices refer to domains other than simply the knowledge of the principles of action

76   *Elements for an Anthropology of Technology*

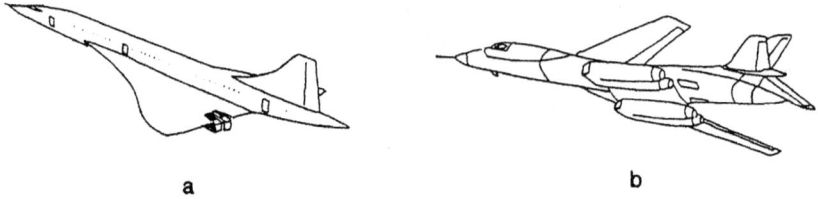

Figure 36. Mach 2 planes. *a*, Concorde; *b*, Rockwell B1B bomber. (Drawing by Ph. Gouin.)

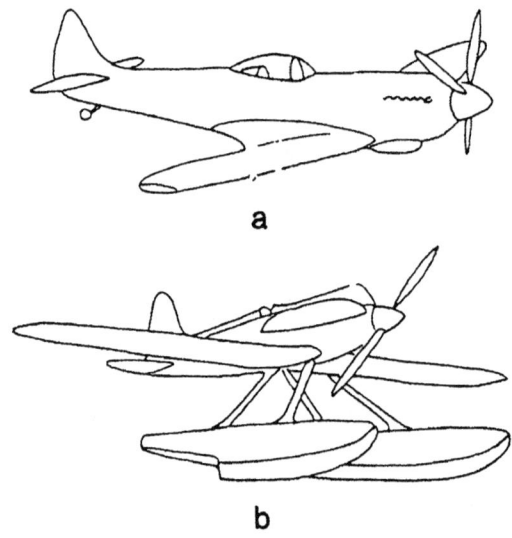

Figure 37. *a*, World War II fighter; *b*, 1930s racing hydroplane.

on the material world. Thus, even in our so-called technical and efficient societies, an engineer's conception of what a given artifact should be is heavily influenced by already existing designs. As Quilici-Pacaud (1989:54) put it, "the first operating artifact freezes the genus, which can be the only right one." It then takes time for the "tendency" to take over and for a new and more advanced technological shape to be developed.

What about other societies? How much do such social representations weigh on the transformation of the technological systems studied by archaeologists and ethnologists? These are questions that I have already raised, but they are ones which have to be asked again and again. The example of modern aeronautics corroborates the existence of forms of technological thought other than those related directly to the physical aspects of action on matter or the stylistic meaning of decoration on artifacts.

Chapter 4

Social Representations of Technologies

TECHNOLOGICAL TRAITS AS SIGNS

According to Bresson (1987:935), in psychology and linguistics one speaks of representation

> when the object on which behavior bears is not the one that gives them their meaning, but a substitute for this object. In order to speak of representation, it is necessary that a relation exists between two systems of objects (real or mental): one being the representative of the other, the represented.

Thus, a mock-up, a map, figures, diagrams, and the sounds used in verbal communication are all representations.

On the other hand, in a very less precise manner, by "social representations" anthropologists mean sets of ideas shared by members of a given social group. These representations can be explicit: "Anga women do not make garden fences because . . ." Or they can be implicit: the organization of a grammar in the head of the speaker, or the sense of kinship structure that precludes a child from calling parents by their first names (Bresson 1987). It is in this same vague way that I have been talking of social representations of technologies among the Anga of New Guinea, or among aeronautical engineers in technologically advanced Western societies.

Psychologists and linguists study the relationships between representations and communication, and their works should have a bearing on our subject, at least some day. My purpose here will be only to delimit with a little more precision the term "representation" as it is used in the anthropology of technology.

It seems to me that "representation," as I intend it, can be employed in three different contexts. First, any technological action implies the existence of totally unconscious mental operations, which I believe are more or less representations. These mental schemes are the kinds of procedures which, for instance, guide the movement of our hands and fingers when making a pot. Anthropologists know very little about them. For example, we know that they cannot be learned through language alone: one cannot explain to somebody over the telephone how to make a net (Bresson 1987:934, 954). This has important consequences, as we shall see later, on the transmission of technological knowledge.

I suggest that it would be of interest to know if each society actually develops only some of the basic mental operations aimed at technological action among those that are potentially shared by all human beings. An important advance then would be to define these basic mental operations. Leroi-Gourhan's attempt at describing and defining different types of percussion is certainly a step toward this end. But we are still far from being able to set down the basic mental schemes involved in, say, knotting, polishing, digging, or compressing. In short, nearly everything remains to be done in this respect. Even transcultural studies of the most simple body techniques (such as rocking babies or carrying loads) are only in their beginnings (Bril 1986).

A second category of technological representations are those already mentioned in Chapter 1 as a part of "specific technological knowledge." These range from know-how that can be expressed ("you have to fit the washer before tightening the nut"), to automatic—but still learned—interpretations of feels, sights, smells, sounds, tastes, and also such kinesic feelings as position in the gravity field. They also comprise classifications of materials, tools, movements, people's roles in technological actions, and results of these actions.

Thus, alongside, and not very far from, these unconscious mental operations and sequences of operations that are aimed at particular elementary actions on the physical world are other mental representations of technologies, which one might call "project schema." These are the mental models which order the articulation of technological elements (matter, tools, gestures) during a given operation, as well as the concatenation of the various operations that are directed at a particular technological result. According to Marx (1976:284), "what distinguishes the worst architect from the best of bees is that the architect builds the cell in his mind before he constructs it in wax." Thus, there are mental schemes for "getting dressed" or "building an Anga house," which serve simultaneously as algorithms which are known from beginning to end, and as set combinations of operations and elements (e.g., "if I do step one, I must then undertake step two"). A jack-of-all-trades may pick up various bits and scraps here and there, because he has such a projection somewhere in mind. In the same way, an Anga man would locate in advance, mark, and protect or gather some liana, bark, pieces of wood, stones, and earth, to be used another day for a specific technological action such as making a hearth.

These particular representations are but a part of other social representations, which might not have much to do with technological actions, but which definitely affect the choice of technological elements as well as the choice of entire technologies by a given society. The Anga cases and the aircraft examples examined in Chapter 3 gave at least some indication of the links between particular technologies, or sets of technologies, and nontechnological social representations.

Influences can go from general social representations to technological ones, or the reverse, or both. Haudricourt's (1962) intuitive and seminal paper, "Domestication des Animaux, Culture des Plantes et Traitment d'Autrui," is an exploration of "both." He suggested that the different ways people take care of animals and plants in the East and in the West might have an effect on the way those in power behave. And conversely, since technological behaviors reflect social order (e.g., "men do this and women do that"), social classifications of animals and plants reflect the social structure. The logic of technological choices and classifications, as observed in a physical action on matter, might also participate in systems of meaning and reference (see below).

The third category of representations concerns the immediate informational content of technological actions; some aspects of technology look very much like the particular representations that we call "symbols." I will return to this issue shortly.

It should be noted that any type of know-how includes unconscious mental operations, which means that the first and second categories of representations discussed above are highly permeable to each other. They only differ by the degree of consciousness of the control involved. For instance, when the potter's hands are shaping a pot, unconscious algorithms set the order of the microactions and movements of fingers, but other mental procedures, also algorithms, that have become automatic but which are nevertheless consciously known exist as well. The same is also true of the second and third categories of representations, since representations of technologies not aimed at direct communication might nevertheless belong to wider systems of meaning, which are basic to the third category. In view of the present state of the question, the approaches used today, and how much is actually understood, I believe that, however heterogeneous these three provisional categories may be, they can still be of service.

Among the various sociological questions that can be asked about representations of technologies, two have received relatively more attention than others. The first is how technological knowledge is shared and transmitted, the subject of current ethnological work which has been published in the journal *Techniques et Culture*. Salmona (1983) has studied the sociopsychological aspects of learning technological knowledge among stock-breeders, gardeners, and arboriculturalists in southern France. Chamoux (1983) has focused analysis of the pattern of transmission of technological knowledge and know-how among Mexican Indians on the difference between "competence" and "performance." And finally, Geistdoerfer (1983) has studied the effect of social stratification on the transmission of technological knowledge among fishermen in Quebec.

Generally speaking, the effort to understand how various types of technological knowledge are shared (or not) in a given society leads to the key problem of the link between technological specialization and social organization. This is important in any society, but becomes a more acute issue when dealing with societies in which it is knowledge, and not tools, or materials, or machines, that make up the bulk of what is needed to perform a technological action. The other question—the most "classical" so far—concerns the interface of function and style, which I have rephrased earlier as the issue of identifying choices that are arbitrary. Before going into this matter further, I would like to come back to something Leroi-Gourhan wrote more than forty years ago.

Leroi-Gourhan's two volumes of *Evolution et Techniques—L'Homme et la Matière* (1943) and *Milieu et Techniques* (1945)—contain data that are still crucial to an anthropology of technological systems, from both a theoretical and a methodological point of view. He defined categories of "elementary action on matter" (percussion, use of fire, water, air, and forces) and indices for measuring the dynamic features of artifacts in order to construct a classification of primitive (or "traditional") technologies. Divided into "transportation," "manufacturing technologies," "acquisition technologies," and "consumption technologies," such classifications and descriptions comprise the bulk of the two books. But it must be stressed that these classifications were not made for their own sake: they were created in order to ask anthropological questions of technologies. Leroi-Gourhan's concern was to identify and understand where and how other social phenomena interface, and interfere, with technological evolution, specifically with innovation and borrowing. For this purpose, he first postulated the existence of a technological determinism "comparable to biological determinism, with as much overlap, as many exceptions, but with as much clarity, in the ensemble" (1945:334), and defined the concepts of "tendency" (*tendance*) and "fact" (*fait*).

By "tendency," Leroi-Gourhan meant that characteristic of technological evolution by which, independent of any direct connection, processes and tools appear that make use of the same forces and exhibit the same mechanical, chemical, and other properties, in response to technological problems posed in identical terms. It is what causes roofs to be peaked, axes to have handles, and arrows to balance at a third of their length from the head (1945:338).

The "fact," he continues,

> as opposed to the tendency, is unforeseeable and particular. It is quite as much the encounter of the tendency and the thousand coincidences of the milieu, i.e., invention, as pure and simple borrowing from another people. It is unique, unextendable, an unstable compromise established between the tendencies and the milieu. [Leroi-Gourhan 1941:28]

Obeying a general tendency, ethnic groups produce objects whose morphology or mechanical properties differ to the degree that the observer is meticulous in observing them. As a result, the facts present "degrees," which correspond to their progressive individualization. In other words, the more the cases expressing (objectifying) the same tendency differ from each other, the more they correspond to particular human subgroups (see below).

For Leroi-Gourhan (1945:336), the external milieu (*milieu extérieur*) provides that which could potentially be used by a given society for its technological action. It comprises geographical, zoological, and botanical features as well as other neighboring societies. The internal milieu (*milieu intérieur*) is made up of the mental traditions of a given human group. A part of the internal milieu is the technological milieu (*milieu technique*), or the mental traditions that more specifically deal with action on matter (1945:340). When in contact with a given mental tradition, the tendency materializes itself in a particualr material culture, or, as Leroi-Gourhan (1945:339, 346) puts it, in a particular "technological group" (*groupe technique*).

The "degrees of the fact" are the steps by which a classification of a given technology becomes more and more detailed. Thus, the first degree of the fact corresponds to the main function of a given technology, and can be identified with the tendency. The tendency, for example, is to use a hammer, harpoon, or spear-thrower (Leroi-Gourhan 1943:34). The subsequent degrees of the fact correspond to secondary physical aspects of the technology in question. The last degrees of the fact correspond to the last branches of a tree diagram. They are those details having little or no physical efficiency and which can be explained by their relations to the internal milieu: they can have technological, religious, and decorative explanations at the same time (1945:342).

Leroi-Gourhan's last degrees of the fact correspond, therefore, to the realm of what is today called style, and his technological milieu is exactly what I am calling representations of technologies. Most of his ideas and conclusions remain basic to a modern anthropology of technological systems, and I shall describe some of them here in order to show that, although they are forty years old and yet to be translated into English, these two books still set the standard for the field.

Leroi-Gourhan demonstrated that the technological milieu is continuous. Technological actions or artifacts have to be related to ones already existing in order to take shape. In other words, a particular technological trait has to link up with or build on other technological traits which already exist. This conclusion has important consequences: at a given time, the technologies of a particular human group are tied together by common underlying technological traditions (Leroi-Gourhan 1945:344–45). Thus, Leroi-Gourhan forty years ago was already pointing to the part played by social representations of technologies in their systemic aspect.

He developed the hypothesis that as a technology evolves, the success of a borrowing depends on its coherence with the internal milieu (Leroi-Gourhan 1945:356–57), and for this purpose he defined the concept of "favorable milieu" (1945:359, 375ff). Invention, too, is a result of the evolution of the internal milieu, and Leroi-Gourhan (1945:376–95) gave us what is still the best, though still far from complete, anthropological account of this crucial phenomenon. (We know that an invention is necessarily linked to the already existing technological system, and we also know a lot about the conditions of adoption of an invention; but we don't know much about the processes which make individuals and groups get out of the routine and invent a new technique). He also studied the effect of technological change on social representation and social organization, taking as an example the introduction of reindeer among the Eskimo in 1890–1900 (1945:364–69).

The relevance today of Leroi-Gourhan's work is clear, and we should keep his contributions in mind when looking at modern rediscoveries of the symbolic dimension of material culture.

STYLE AND FUNCTION

We have just seen that there are two types of technological traits. One type of trait communicates information, such as the decorations found on an artifact or secondary features of shape having few or no physical functions. The other type is designed to have an action on the material world, though the features of it may show some degree of arbitrariness from the point of view of its physical action. For example, the use of a given material that cannot be recognized after processing, but which was chosen as appropriate because it belonged to a wider nontechnological social classification of the material world.

In the former case, the color or shape of an artifact, or the particular design of its decorations, is read by those who look at it and is correlated with a particular social identity based on gender, age, social stratification, membership in a particular ethnic group, and so on, or with particular social events, perhaps weekdays, or Sundays, or ceremonies. This reading can be done consciously: for instance, people know that admirals have a given number of stars on their collar. But most often this "reading" is unconscious: I "know" that a particular nonchalant pedestrian I am looking at in Paris is an Englishman without analyzing whether his moustache, tweed jacket, or parting of the hair tells me that he cannot be anything but an Englishman. An admiral's or an Englishman's clothes voluntarily announce him as an admiral or as an Englishman.

However, in other cases, and in fact very often, the message is involuntarily sent and, nevertheless, is read as information on social identity. Thus, a type of car at a point in time can announce a 30- to 35-year-old businessman with a six-digit income, who is likely to be a sportsman with no children. It is also true that ten years later the same car, quite rusty by now, may say that its present owner is poor and doesn't place much value on status conferred by automobiles. Sometimes, false messages are purposely sent, and it may be difficult to distinguish the real punk or executive from someone who occasionally wants to convey that image. Bromberger's (1979) programmatic paper provides theoretical and methodological suggestions for a "semio-technology" that may help anthropologists investigate and understand the processes underlying these particular choices.

For those technological traits that are primarily aimed at an action on the physical world, things are even more complicated, because the social choices from which they stem may be totally unconscious. In the case of the use or non-use of a particular material, the unconsciousness may reside at different levels: for instance, the association of a given material to a gender, but also the very choice of the material for a particular technological action, and the logical relation of aspects of this material with aspects of other materials. These may all be—and very often are—the result of unconsciously acting on mental representations. It is the ethnologist's or archaeologist's task to reconstruct the complex—indeed very complex—processes by which a particular technological trait, a vegetal material, for instance, takes on its "meaning." But this will prove to be a very difficult task, one which cannot be accomplished in any simple or direct manner, and one which is not likely to lead to quick generalizations. We must be aware of the complexity of systems of meaning when looking critically at symbolic interpretations of material culture.

When dealing with a system of meaning, some basic questions have to be asked concerning its mode of functioning and its role. According to Beneveniste (1974:57), looking at the mode of functioning of a system of meaning would require "identifying the units that it brings into play to produce 'meaning' and specifying the nature of the 'meaning' produced." Just identifying these units is already a tricky proposition, as the line between stylistic (communicative) and functional (linked to physical action) is hard to draw. It is quite easy to point to stars on an admiral's collar, but what about a cassowary trap hidden in the thickest part of a primary forest in New Guinea? Is it the use of a snare, as opposed to a dead-fall trap, that works as a sign, as an objective indicator of meaning? Or rather is it some part of the snare or trap, or even the technological principles brought into its design, that serves as a sign?

To specify the "meaning" produced, it is first necessary to identify the mode through which the system of meaning operates in the material culture. Since sight is likely to be the main sense used, this leads to the important issue of identifying who are the people who may look, or who have looked, at particular technological features that we assume have a role to play in a system of meaning. And, first of all, are these features visible? Obviously, details of a lithic industry that can be observed only with the use of a binocular microscope had no meaning for those people who made and used the objects, although they may have some interesting "meaning" for archaeologists. In other words, the question of the *domain of validity* of a given system of signs also has to be raised: for whom is it designed, in time as well as in space? Only then can one satisfactorily investigate the relations that link the different signs and give them their relative meaning (Beneveniste 1974:57).

The answers to these basic questions would tell us who are the people involved in using a given system of meaning based on features of material culture, and, if we succeed, how it works. Another question remains: what is the *function* of this particular system of meaning? We can more or less imagine why people talk, write, or look at traffic lights. But for what particular purpose do parts of material culture belong to systems of meaning? This is less obvious. Fortunately, we can rely here on a direction which, in a way, is one of the guidelines of anthropology: we can assume that one of the functions of this particular system is to express social differences. On the one hand, technological choices may express differences between groups, participating in what Lévi-Strauss (1985:xiv) called the wish of "each culture . . . to oppose those which are around it, to distinguish it from them, in other words to be oneself." On the other hand, technological choices may also express within-group social differences in terms of age, gender, and other criteria, as I have already indicated.

A subsidiary issue, but a very important one, has to be raised here. To say that, in a particular context, some features express social identity, does not tell us *why* social identity is expressed in this way and not in another. More specifically, that material culture may be used in what some people call "social negotiation" or "manipulation" has yet to be explained. If such behaviors exist (which seems obvious, as shown for example in the case of political demonstrators who dress so as to express their opposition to a given idea), their range and frequency in various societies nevertheless still has to be specified. Furthermore, individual behaviors have to be distinguished from group behavior, for fundamental epistemological reasons (see Durkheim 1950; or Devereux 1956). In other words, it is one thing to ascertain that women and men do not dress alike, but another to demonstrate that women, as a group, oppose men, or younger women oppose elder women through their clothing. Taking the example a step further, to postulate—as does Hodder (1982, his Chapter 8) for instance—that an individual purposely manipulates pieces of material culture in order to express some personal view on the way society works is even more difficult to prove. Once again, one cannot get rid of such fundamental axioms of sociology as the postulate of nonintentionality in social relations, or the methodological caveat on mixing psychological and sociological phenomena.

Even when it is demonstrated that a given social group, or individual, voluntarily uses material culture to express particular social relations, including conflicting relations, we are still far away from being able to make generalizations about material culture and social relations.

The problem here is to explain why one given aspect of material culture (e.g., the shape of pots), and not another (e.g., the way people walk, dress, or hunt) is used to express certain social relations, and thus, we come back to an earlier question: how does a particular variation in material culture fit into a particular system of meaning? Without explaining how the context (namely, the particular social representations held by a human group) determines the choice of a particular technological trait to express some social relation, no generalization is possible. This unsolved problem, by the way, considerably reduces our capacity to interpret archaeological remains.

In addition, many technological traits that show some arbitrariness from the point of view of their physical action on matter may well be involved in systems of meaning which are not primarily aimed at communication. A design on a pot may "say" something about the ethnic identity of its maker or user; it may even say something about conflicting relations. But it may also say nothing of the kind, and merely express an aesthetic feature. In this case, it would tell archaeologists and ethnologists only whether a given type of artifact lends itself to expressing aesthetic feelings, if indeed they are able to differentiate aesthetic decorations from those carrying other kinds of information. But, in any case, it would not communicate much to the people who use it, except that it belongs to the category of, for instance, "nice" objects. As for technological traits that obviously do not carry any immediately readable meaning (as, for example, the cassowary trap already mentioned), they might participate in systems of meaning that have nothing to do with social identity or social stratification, but instead with general questions on the meaning of the world.

Although technological traits not immediately bearing information are sometimes part of systems of meaning related to social differentiation and hierarchies, or to social relations of production, it would be very difficult—in fact, almost impossible—to infer comparable social relations in another setting just on the basis of the presence of similar technological traits, because the complete context in the new setting will be missing. For instance, the fact that the use of a given species of plant or animal by the Anga may be related to a basic opposition between men and women (see Chapter 5) does not allow us to infer that the differential use of similar species in another society also has something to do with male/female relationships.

To sum up, the complexity of the involvement of material culture in systems of meaning is such that we probably need to begin afresh, gathering new data to which we explicitly address this difficult question: how are style and function interdigitated? Or put another way, which are the technological traits that are subject to possible technological choices? These questions, as well as a thorough examination of the mode of functioning, and role of systems of meaning, involving material culture are particularly critical. I do not see how, without first finding answers to these basic questions, we can claim too know the meaning of technologies, other than *ad hoc* conjectures and plausible but untested hypotheses. With these warnings in mind, we may now review the issue of style in anthropology.

STYLE WITHOUT FUNCTION?

In archaeology as in ethnology, the question of style has been raised with respect only to artifacts. This ignores the variability in other elements of technological action.

Style has two basic meanings. The first is a diagnostic one, in which style is a tag attached to a culture as a whole (see Kroeber's [1957] *Style and Civilization,* for instance), or refers to an aesthetic aspect of a culture. In these terms, style has no explicative aspect.

The other approach is to consider that style carries information on social identity. This view has been developed by Wobst (1977) and Wiessner (1984), for example, to whom I will return in a moment. But let us first consider style as defined by Sackett (1982):

1) *isochrestic* style refers to ranges of shapes or forms that are adapted, with equivalent efficacy, to a given (physical) function or aim;

2) the craftsperson makes choices in these ranges;

3) the probability is small of finding similar combinations of choices in two different societies;

4) these choices are socially transmitted.

These points are of great interest for an anthropology of technology, but, unfortunately, the use that Sackett makes of them is disappointing. Indeed, Sackett's "isochrestic" style seems to have no function, neither physical (by definition) nor informational, because, while it expresses an ethnic "mark" resulting from the social and historical context in which an artifact has been produced, this information is addressed *to the archaeologist alone*. The only function attributable to Sackett's isochrestic style can be deduced from an observation he makes on a possible adaptive advantage of particular shapes. I interpret his remarks as follows: shapes with identical efficacy at a given time may, in the course of subsequent transformation, give birth to new shapes the efficacy of which is no longer identical. Thus, isochrestic style could lead to differential adaptive advantages. Although this does not tell us anything about the question of style as a means for communicating social information, this observation is of importance from an historical evolutionary point of view.

Interestingly, Sackett (1982:105) also mentions that "knapping techniques for reducing cores and producing tool blanks, alternative types of marginal retouch and burin spalling, and varying angles and wear patterns," as well as "hunting strategies, butchering practices, and the other activities pursued by a given prehistoric society," may in some way relate to isochrestic style. But unfortunately, his idea that entire techniques may also show isochrestic variations, an idea with which I fully concur, is not developed or demonstrated.

Another interesting idea, but also one that Sackett fails to develop, is that the same phenomenon, but seen from different points of view, may be both stylistic and functional. From the functional perspective, one would investigate how a shape and a physical result are interrelated, while from the stylistic side one would focus on the choices made by people among the various shapes or forms with which they were confronted whose physical efficacy was identical. For instance, if an archaeologist studying World War I remains is able to distinguish between German and French helmets and gas masks, as well as between bottles of Schnapps and *gnole*, and if he or she digs up shelters, officers' messes, and aid-posts, he or she will identify two populations, whereas an archaeologist unaware of isochrestic variations would identify three populations, living respectively in shelters, officers' messes, and aid-posts (Sackett 1982:77–78).

Although these questions might illuminate archaeological problems, they are of little interest to ethnologists, first because an ethnologist would normally know the origin of the artifacts he or she is looking at and, secondly, because Sackett does not ask the crucial question concerning the differential *efficacy* of the artifacts in question: Mauser and Lebel rifles, for instance. In short, from the point of view of an anthropology of technology, Sackett's approach lacks the references to the social representations of technology that might help us to investigate the function of isochrestic style in a system of meaning, or that might allow us to examine the social basis for isochrestic choices. Finally, Sackett does not explore the relation between specific shapes and particular physical results, as for example for the different types of helmets, gas masks, and rifles. In this respect, Sackett asks fundamental questions, but in my opinion does not answer them.

TECHNOLOGY AND INFORMATION

Wobst's (1977) work on Yugoslavian folk costumes is representative of the best analyses of the informational content of material culture as done by anthropologists. In particular, he is the one who first clearly drew attention to the fact that material culture has functions that relate to the exchange of information as well as to matter and energy. His paper shows how particular items of material culture—parts of costumes—can express different social identities of the wearer, according to the distance at which these differences can be perceived (1977:328–29, 337). He also hypothesizes that information expressed in folk costumes plays a role in boundary maintenance, a theme developed further by Hodder (1982). In particular, Wobst emphasizes the importance that those traits used to express social identity be visible, and suggests that only a few artifacts are suitable for such a function.

Here again, the problem is that there may be many ways to express social identity, other than by wearing visible traits of material culture. Certainly distance plays a role, and it is true that language or postures can efficiently communicate social identity only over a few meters. But communication through material culture cannot be reduced to one context of encounters between members of different social groups, even though this fits well with some classical questions in archaeology. Hopefully, communication between foreign groups is not solely a matter of borders. And many groups do not use material culture for this purpose. For example, the Ipmani and Baruya of the

Eastern Highlands province of Papua New Guinea look alike, share the same material culture, and speak the same language; nevertheless, they have been fighting and killing each other for years in boundary disputes. Thus, so long as Wobst (or Hodder, see below) does not tell us why some ethnic groups mark their ethnic identity with particular items, and why others do not, any inference that archaeological artifacts might also have functioned to maintain borders remains purely speculative.

Furthermore, Wobst's approach says very little about the use of material culture within a given social group to express gender, age, status, and so forth among people who may share common interests rather than oppose each other. Nor is mention made of the involvement of material culture in systems of meaning other than pointing out immediate (and visible) symbolic units, which, as I suggested earlier, may be more important for the evolution of technological systems than the stylistic marking of social identity. Finally, although Wobst's paper is an important contribution, it makes no mention of Bogatyrev's (1971) pioneering work in the 1940s, *The Functions of Folk Costume in Moravian Slovakia*.

According to Ogibenin (1971), Bogatyrev was already interested in costumes as "signs" in the mid-1930s, influenced by the functionalist linguistics of the Prague School which emphasized the functional (differential) role of sounds in language. Bogatyrev (1971:81) suggested that an object—for example, a power hammer—becomes a sign by referring to something that is beyond itself (Bogatyrev 1971:81). Concerning costumes in particular, he was interested in the conditions under which a protective item becomes a link between the wearer and particular social groups (Ogibenin 1971:14). Pieces of costume can be identified with signs because those elements of clothing that are given meaning are conventional. In a given context, one learns to read these particular signs (Bogatyrev 1971:84). In the case of Moravian costume, Bogatyrev believed that wearing the right costume in the right circumstances was compulsory, and done without any conscious control by the individual (Ogibenin 1971:20). This conclusion, of course, came as a direct consequence of the linguistic analogy.

For Bogatyrev (1971:82), a costume is both an "object" and a "sign," and the informational function of a costume is secondary. He recognizes many meaning "functions" in costume, each being an expression of a particular social behavior of the wearer. Thus, in Moravian Slovakia, "functions" of costume denote occupation, wealth, status, and regional and national identity. They may signal the erotic, the esthetic, the magical, and of course they may be practical (Bogatyrev

1971:43–76). The hierarchy of these functions varies according to the context in which the costume is worn. In a transition from everyday to holiday costume, Bogatyrev suggests—but unfortunately does not explain how he obtains this result—that this hierarchy would change as follows (Table 6):

The everyday costume is first of all an object, while the holiday costume is first of all a sign (Bogatyrev 1971:91). According to how it is combined with other elements in a particular context, a piece of costume—a unit in a system of meaning—can be involved in various functions. Thus, when associated with hemp trousers, the *vonica* (a hat decorated with flowers) indicates that the wearer is a recruit. Combined with trousers that are not made of hemp, it means that the wearer is a groom (Bogatyrev 1971:41–42).

Finally, Bogatyrev (1971:102–5) recommends the possible use of the "structural-functional method in the study of village buildings, farm implements and other items of material culture, as well as in folklore (magic, folk tales, songs, incantations, etc.)." What Propp (1968) and Lévi-Strauss have done with a part of this program is, of course, well known.

To date, Bogatyrev's work contains the fullest exploration of the mode of functioning of a material-culture–based system of meaning. For instance, he made explicit how a given unit of costume acquires its meaning through its position in a particular combination of units.[7] What is missing in Bogatyrev's, as well as in Wobst's, approach is an account of the implicit relationships between particular features of units of costume and other elements of other systems of meaning. Besides the links it has with other pieces of costume, a *vonica*, vest, or trousers may partially take its "meaning" from its shape, color, material, or decoration in relation to other shapes, colors, materials, or decorations. This aspect, too, has to be taken fully into account in a study of costumes as signs.

Wiessner (1982, 1983, 1984, 1989) has done ethnological—or ethnoarchaeological—fieldwork among the !Kung San. She considers that there is a behavioral basis for many, though not all, of those variations in material culture that have been called "style" (in Wobst's terms) in archaeology (1984:193). But she also considers that style does not simply carry information on ethnic identity; it is also an active tool used in social strategies (1984:194). She then asks the crucial question: in what circumstances are artifacts used in these "social strategies?" Her fieldwork and data focus on artifacts on which people make comments, that is, compare the way they make and decorate them (e.g., arrowheads [1983] and headbands [1984]). She listened to these comments and asked questions of the artisans.

TABLE 6
Functional Hierarchy of Moravian Folk Costumes

Everyday	Holiday or Ceremonial	Ritual
1. practical	1. holiday or ceremonial	1. ritual
2. social status or class identification	2. aesthetic	2. holiday
3. aesthetic	3. ritual	3. aesthetic
4. regionalistic	4. nationalistic or regionalistic	4. nationalistic or regionalistic
	5. social status or class identification	5. social status or class identification
	6. practical	6. practical

Source: Bogatyrev 1971:43–44.

Arrows are used in *hxaro* exchanges (see Wiessner 1982). These take place between individuals, half of whom can usually trace consanguinal ties. These *hxaro* relationships take the shape of help, sharing, and visits. Exchanged items can be anything that is not edible (1982:70). Arrows are visible and exchanged. They are made by the hunter himself in 57% of the cases, come from a partner one to twenty kilometers away in 26% of the cases, and from partners farther away (up to 200 km) for the remaining 17%. The !Kung are able to comment on 3 mm variations in width and on 2 mm indentations, and on the shape of arrowheads. They say that arrows clearly indicate the identity of the hunter as well as the sharing—or nonsharing—of identical "values" concerning hunting, territory, and general behavior.

Wiessner's (1983:269) studies show that:

1) !Kung arrows differ from those of the !Xo and G/wi (two other groups 300 km away);

2) !Kung "reacted to the G/wi and !Xo arrows with surprise and anxiety";

3) arrows of G/wi and !Xo, who are in contact during the dry season, differ less from each other than from those of the !Kung;

4) !Xo and G/wi declared that !Kung arrows "were pathetic and could not kill anything."

Wiessner thus suggests that arrowheads are used as an ethnic marker in the Kalahari. Nevertheless, as the !Kung are separated from the two other groups by 300 km and having never seen their arrows before, it seems safer to suggest only that these arrows might have *been* used as ethnic markers. Once again, it is very important to ask who is likely to look at an artifact to which we attribute a position in a system of meaning. Wiessner also says, very wisely, that considering only one kind of artifact does not reveal enough to understand the general use of material culture in a given society.

Headbands are made of glass (formerly ostrich egg shell) beads, and are made, worn, and exchanged by women (Wiessner 1984). One individual keeps them for a period of time that ranges between two months and two years. Women do compare and comment on the design of their decorations. Contrary to what archaeologists might think, the same decorative design can be found in different linguistic groups, and the differences observed between adjacent groups are more important than those observed in groups 400 km apart. This observation, incidentally, might point in the direction predicted by Hodder (1982), although Wiessner (1984:217) argues against this possibility. Headbands participate in "strengthening relationships of loose but generally positive affiliation with kindred members and affinal kin, and ... in negotiating individual's identities and carving out individual niches" (Wiessner 1984:210). It should also be noted that, according to her, "negotiating identity relations" is done "consciously or unconsciously" (Wiessner 1984:209). Variability seems to correlate with the frequency of the comparisons made by the people.

The most important point here is that !Kung women themselves say that they make beautiful artifacts on purpose, "to impress the opposite sex, ... to promote reciprocal relations, ... to gain self-satisfaction, ... and to impress Bantu agriculturalists" (Wiessner 1984:204). For once, voluntary social interaction through material culture items is an unquestionable fact. Nevertheless, Wiessner, again very wisely, insists that

> for items of material culture which do play an active role in social relationships, identification through comparison provides a basis for the assumption commonly made by archaeologists that style can provide a measure of interaction. . . . However, it suggests that style can only do so within the realm of those who are defined as comparable, under certain conditions, and in the context of certain types of relationships. The contexts in which stylistic variation over space can be used as a measure of interaction need to be further explored. [1984:229–30]

We shall soon see how some archaeologists (and Wiessner herself) sometimes forget that such material-culture–based interactions remain unpredictable. We must also keep in mind the range and kind of "social negotiations" under discussion here: namely, impressing other people. Unfortunately, to the crucial question of why one artifact is used to mark ethnic identity while another is used to impress people, Wiessner (1984:228) gives a circular answer, which does not explain anything. Style, for Wiessner at least, has a function, and the questions she asks at the end of her paper, in order to further develop a "behavioral basis" for style, are fundamental ones.

Finally, it should also be remembered that material culture is not only made of artifacts, and that its involvement in systems of meaning goes far beyond what has been described by Wobst, Bogatyrev, or Wiessner. Furthermore, none of these authors seems to have the slightest idea of what is perhaps the most important feature of material culture: its systemic aspect. In Wiessner's case, this may be due to a quirk which gives ethnoarchaeology its *raison d'etre*, while also defining its limits: it is *archaeological* questions that are brought to ethnological fieldwork, instead of general anthropological ones.

EXCESSES OF THE CAMBRIDGE SCHOOL

The Cambridge School of symbolic archaeology is the name given to the team of researchers who have gathered around Ian Hodder in recent years and devoted their efforts to a consideration of what has been called "symbolic" and "structural" archaeology. "Excess," here, simply means what it means in Webster's dictionary. The Cambridge School has extended the traditional limits of the study of material culture by archaeologists, both by widening its scope (especially by doing intensive ethnoarchaeological fieldwork), and by trying to ask of sherds, beads, and post-holes some very classical questions of social anthropology. Such a revival of anthropological interest in material culture has no parallel, except among Cresswell's CNRS team of ethnologists in Paris, "Techniques et Culture," which has been working together since the early 1970s, and more recently at the Centre de Sociologie de l'Innovation (CSI) of the Ecole Nationale des Mines, also in Paris.[8] At a time when entire technological systems are disappearing as fast as the societies that developed them, such efforts to do so-called ethnoarchaeological fieldwork and to put the study of material culture back into the realm of social anthropology are priceless. In this respect, it would be disastrous to limit the range of this kind of investigation by limiting either its data or its theoretical framework.

Here, I shall assume that the work of Hodder and his colleagues is familiar territory, and I shall comment on it only briefly. Criticism is not only the best form of praise, it is also a way to call for an anthropological study of material culture that would encompass both the reconstruction of technological systems from archaeological remains and the investigation of living material cultures.

Paradoxically, one of the outcomes of the endeavors of the Cambridge School to expand the domain of the archaeological study of material culture has been to narrow excessively the theoretical approach to technological systems. For this reason, my main criticism will not be to argue that one or the other of Hodder's positions are wrong in themselves, but rather that they are far from being the only anthropological statements that need to be made about material culture.[9]

To be brief, symbolic and structural archaeologists have a tendency to limit: (1) the data of their studies to artifacts, (2) the analysis of these objects to their (de facto) immediate informational content, and (3) the study of this information content to the role it may play in conflicting social relations. I consider this narrowing of already narrow aspects of material culture to be incompatible with what should be the common target of both archaeologists and ethnologists in studying material culture: relating the structural features of technological systems to other social relations.

Whether they are stools, pots, spears, headbands, British cemeteries, or American gardens, artifacts are just artifacts. Being parts of systems, they can only be understood if they are related to the other elements of the system (or systems) they belong to. The information that must be taken into account includes not just the shape or decoration of the artifact, but also, as already mentioned, the materials it has been made from, the way it has been made, the way it has been used, the artifacts or technological behavior that could have been made or used instead, and so forth.

By looking only at the shape and decoration of an artifact, one misses entire sets of social phenomena that are related to the making and use of items of material culture. Sherds, post-holes, and broken spears can say a lot of things to archaeologists about the materials, techniques, and knowledge involved in their making and use, as well as their relations with other items of material culture and technological behaviors. It is not even reasonable to argue, for instance, that since archaeologists deal mainly with decorations on pots, ethnoarchaeologists or symbolic archaeologists should also deal only with decorations on pots. The study of a living material culture provides a unique opportunity to investigate many aspects of technologies that lie behind archaeological remains. Ethnoarchaeologists are ready to ask women what they think of their headband designs—which is definitely a necessary step in the ethnological investigation of the subject—even though headbands rarely speak to archaeologists. Why not also ask the people how they make and use their pots, spears, or stools? Why not look at the way they carry loads, lie on the ground, hunt, or feed babies? Or, best of all, why not watch and ask questions about how they make and use headbands, *and* carry loads, *and* make and use spears, and so on? Why throw away the systemic aspects of any artifact, or technological behavior, when it is available to investigation by the ethnoarchaeologist? The more that technological relations are dealt with in the observation and analysis of a living material culture, the better our chances of constructing an accurate interpretation of archaeological remains.

Limiting oneself to "style" seems to me an unfortunate way to return to nineteenth-century archaeology, when studying style was the main tool for constructing typologies for their own sake. In a recent paper, Tilley (1986) sets out very clearly the reasons why the interpretation of material culture should focus only on its informational dimension (usually its most immediate or obvious informational dimension). Rightly or wrongly, I shall consider him the spokesperson for other symbolic archaeologists who simply forgot to justify their exclusion of those other aspects of material culture generally referred to as "function" or, as Tilley suggests, "technology"; that is, those features that are aimed mainly at actions on the physical world. To argue, as I have earlier, that it is impossible to tell where function ends and "style" begins is to point to the basic question of an anthropology of technology—the problem of the relation between the physical function of material culture and the social (symbolic) representations of which it is a materialization. In no case is it justifiable to simply forget about the physical functions of technologies.

In another argument, Tilley almost seems to be joking when he says that

> ascribing any specific function to an object is in many, if not all cases, an extremely dubious exercise. A chair may be to sit on, it nominally fulfills this function, but chairs can also be used for standing on or knocking people over the head with or as rulers, pendulums or almost anything else. [1986:2–3]

In other words, the variety of uses of a chair—the range of which everybody agrees goes far beyond sitting—makes it impossible to take its physical function into account. This does not make sense: in any culture where chairs are to be found, their use is limited, and limited by the culture itself. As a functional artifact, a chair cannot be just anything in a culture, and to attribute one or several main functions to a "chair" should not be beyond the abilities of an archaeologist, even if the individual concerned is a symbolic one.

In yet another argument, Tilley is no more convincing than in his previous one when he states that

> specifying a function tells us virtually nothing about the specificity of individual chairs and all the multitude of different chair forms, past or present, their shapes, decorative features, arrangements in different rooms or different types of rooms. [1986:2]

This is simply not true. As a physical object, a chair is the result of other technological actions and of the use of other artifacts, the combined function of which has been to produce it; it is also itself a means of action on the material world. These are two reasons why a chair tells us many things, which are all information on this artifact as a social production as well as an artifact that is socially used. In other words, a chair can provide precisely the kind of information which, as far as I know, seems basic to any anthropological approach to material culture. In effect, a chair, and the way it is made, and the ways it is used (thrown through the windows or burnt as firewood), and the way people sit on it—all of which are actions on the material world—give us fundamental information on the chair as a materialization of social thought: on the place of this artifact with respect to other objects; on the system of materials, movements, tools, and knowledge involved in its making; on the body techniques of its users; and so forth.

For these reasons, I still do not see why the functional aspects of material culture should be left aside, and why its anthropological study should focus only on its stylistic aspects. Not only must both of these aspects be considered, but, as I have tried to explain all along, it is the consideration of both aspects at the same time that provides the wider views of material culture as part of systems of meaning.

Turning now to the study of the informational content that Tilley recommends, other limitations become evident. For him (1986:1) material culture, from the point of view of style, is "a means of cementing together social groups or symbolizing group identity." Indeed, style obviously has such functions, but it does not only have these functions. Furthermore, these same functions suggested by Tilley (1986:11, 13) cannot be reduced—as he suggests—to the communication of information linked to power.

The sharing of a common technological system certainly contributes, in some way, to group solidarity. But the dimensions of meaning that are involved cannot be limited just to the marking of group identity (whatever the nature of the group in question), nor to the "mirroring" of message exchanges, the latter being messages about power relations, if I understand Tilley correctly. In a Marxist view, which I adopt here, social representations of technology—that is, representations of the physical components and aspects of material culture, not just representations underlying features of shape or decoration that immediately communicate something to people able to read them—certainly play a crucial role as a mediator between forces of production and social relations of production. But it is another form of reductionism to suggest that every social relation or representation has something to do with the relations of production, even if they are egalitarian ones.

As already mentioned, I suggest the broader hypothesis that some social representations of technologies—that is, the classification of materials, tools, actors, operational sequences, and so on—definitely are linked to other social representations (of life, death, the supernatural, relations between male and female, wild and domesticated, etc.), and to the evolution of technological systems.[10] Tilley makes a double reduction: he reduces social representations concerning material culture to only those "materialized" in patterns of shape and decoration that are immediately visible (i.e., "style," for most archaeologists); he also reduces material culture, as a system of meaning, to a means of exchanging messages on power relations. It seems to me that these two reductions take us further from understanding material culture as a social production, simply because, in so doing, entire social dimensions of technological systems are ignored.

One important point on which one should agree totally with Tilley (1986:5–7) is his reaffirmation of the "primacy of the social over the individual." Strangely enough, it seems to me that this position goes against what Hodder and other "symbolic" archaeologists have previously suggested. Indeed, the role attributed to individual action in most of their studies would be another central target of my criticism. Mixing the study of the individual with that of the group is as risky as forgetting the nonintentionality of many social phenomena. A recent paper by Yengoyan (1986:370–72) stresses the dangers in anthropology of focusing either on "action theories of behavior in which the individual is the central entity," or on the particular (I add "particular") form of Marxism that is concerned with "relationships between ideology and culture." In the latter, "culture thus becomes the trappings and mystifications that conceal and invert a basic inequality running throughout the society." Both of these observations would be relevant to a criticism of "symbolic" archaeologists, by the way (cf. Hodder 1982:76, 85, for example).

Symbols in Action (Hodder 1982) and a recent paper by Wiessner (1989) are good examples of the excesses to which the step-by-step combination of a behaviorist point of view with a narrow "stylistic" approach can lead the "symbolic" (ethno-)archaeologists. There is no need for more comment on the limitations of this approach to: (1) artifacts, (2) a few artifacts, (3) the immediate informational content of the shape or decoration of these few artifacts. I have just spoken of this, and a glance at any of the books and papers dealing with "symbolic and structural archaeology" will confirm my point. Now, what about the active role of individuals in their culture? A few quotations will demonstrate that if words mean anything at all, people either do not use the right ones or they definitely have a tendency to reduce the social aspects of technological systems to the voluntary use of shapes and decorations of artifacts for modifying aspects of the social organization in which they live. Hodder, for example, says:

> ... to understand what symbolic meaning is given to the artifact types within the particular Baringo context, and to understand how that symbolism is manipulated as part of the social strategies of individuals and groups within the different tribes. [1982:58]
>
> The spear symbolism is manipulated by the young men in their relations with the old. [1982:68]
>
> The calabashes, then, are appropriate for use as a medium for silent discourse between women.... Women use calabashes to disrupt the boundaries in opposition to the older men and to form their local independence. [1982:68]
>
> Younger men and women utilize material symbols which disrupt the dominant social order. [1982:73]
>
> There are thus considerable pressures on young women and [a young woman] uses her dress as a way of expressing her willingness to be incorporated. [1982:81]

If "to manipulate," "to appropriate," "to use ... for," and "social strategies of individuals" are expressions that mean anything, then

Hodder at least gives the impression that individuals may change their society through the medium of style in their material culture.

Things are less clear in Wiessner's papers. She says, for example, that

> ... style is not just a means of transmitting information about identity, but an active tool used in social strategies, because in the process of presenting information about similarities and differences, it can reproduce, disrupt, alter, or create social relationships. [1984:194]

> ... knowing the context of the occasion and the image that they would like to project, they choose a style that would communicate relative identity, whether consciously or unconsciously. [1989:57]

Wiessner appears to be giving an account of the "negotiation" of individual personality, in which case her approach is of great interest for the psychological study of the relations between individuals and material culture. But, since she provides no theoretical link between this psychological approach and an anthropological (sociological) study of material culture, I cannot see what it has to do with the general question of the relations between technological systems, as social productions, and other social phenomena.

On the other hand, if style is "an active tool used in social strategies," then Wiessner shares with Hodder the credo that individuals may change certain aspects of their society through their stylistic decisions. The problem here is not that this may not be the case: from the assassination of political figures to the invention of the steam engine, individual decision obviously plays a role in social transformation. The issue is that any such individual decision has a social context—as Tilley agrees—and, while individuals can certainly make voluntary changes in material culture, the unintentional logic of social relations nevertheless remains largely beyond the reach of individuals. Once again, no one can argue that what Hodder or Wiessner say may not sometimes be true, in particular contexts. But to reduce the study of material culture as a system of meaning to these marginal cases is a dead-end.

A final important point is that, in spite of the limits of their data and theoretical framework, of the context-specific nature of their results, and of the ethnological counter-examples that can be offered for many of their conclusions, symbolic archaeologists have a clear tendency to consider as ironclad laws what are, at most, interesting hypotheses. The case of style and boundary maintenance discussed earlier exemplifies this tendency. Similarly, merely because Kalahari women want to impress other people through the patterns of their headbands, to assume that any stylistic feature has a behavioral basis leads to erroneous results, to say nothing of the temptation to reduce the social aspects of material culture to style, and style to individual behavior.

Chapter 5

Conclusions

Toward a Study of Social Representations of Technology

BACK TO THE ANGA CASE

It may seem strange that the concluding chapter following a series of programmatic chapters itself turns out to be programmatic. Nevertheless, this is the case. Whether dealing with the observation, description, or analysis of technological phenomena, or with the theoretical framework of such studies, each point, so far, has raised as many questions as it has provided answers. Furthermore, in most of the cases, the number of studies already available is extremely limited. Such is the state of the issue.

For various reasons, the anthropology of technological systems, as defined here, is still in its early stages, and is certainly the least developed of all existing subdisciplines of anthropology. Data are even more scarce than methodological or theoretical works on the subject, and a corpus of technological descriptions, the gathering of which would reasonably seem to be one of the first steps of any discipline, is almost entirely lacking.

Given these circumstances, attempting a sociological study of material culture using overly sophisticated theories or methodologies would be a waste of time, and worse, it might give the impression that we know far more than we actually do.

The Anga material underscores the kind of dilemma that can arise from the study of social representations of technological actions. For example, Chapter 3 attempted to show that nontechnological factors play a role in the way the Anga mentally order their material culture. Social representations of elements of the technological system are involved here, and they include representations that deal directly with the physical aspects of actions on matter, such as those designed to trap game, kill enemies, or protect one's body from cold, as well as representations that have nothing to do with any sort of physical necessity, such as those concerning the decorations on the shaft of an arrow. The hypothesis has been made that these particular social representations might be part of a system of meaning involving other social phenomena that are not exclusively aimed at physical actions on the material world. This particular system of meaning determines, at least in part, those technological choices that prove to be physically arbitrary.

I have assumed that various components of several Anga technologies, represented here by artifacts and products as well as by technological processes (e.g., house-building, gardening, bow- and arrow-making and use, use of bark capes, salt-making), might be elements of a system of meaning that encompasses all of them. To find the appropriate units, the mode of functioning, the domain of validity—in other words, to find the logic of this system—remains an unsolved problem. I maintain, nevertheless, that it is an informative and potentially productive strategy to investigate the logic of *each* set of technological choices, one at a time, technology by technology. Here, as an example, I present the results of studying various aspects of the use of pounded bark among the Anga (Lemonnier 1984b).

The Anga use several species of trees and shrubs to make three kinds of intermediate products: bark-cloth, lamellas, and fiber-strings. Bark-cloth is used for capes, semi-skirts (Fig. 38), and "mattresses." Fiber-strings are used mainly to make carrying nets and as ties in pieces of clothing, body decorations, and so forth.

Several species are gathered in the forest and/or cultivated. These are mainly: *Ficus elastica* Roxb.; another unidentified *Ficus* sp. (here referred to as "midzamanga," its Baruya name); *Broussoneta papyrifera*; and one Urticaceae.

The operational sequence of the processing of bark comprises the following steps (see Table 7): stripping the bark from the tree, peeling the epidermis, scraping the inner bark, pounding with wood and/or

a. skirt   b. bark cape

Figure 38. Anga bark cape and skirt, New Guinea (after Lemonnier 1984b).

stone beaters (Fig. 39). The result is then torn into lamellas and fiber-strings or made into capes and skirts.

From one linguistic group to another (there are twelve linguisitic groups in all), there is variability in the species used, the way they are obtained (wild or cultivated), the kinds of artifacts made with them, the sexual division of labor in the processing of bark as well as in the making of particular artifacts, the kind of pounding, the type of bark used for a particular artifact, the differential use of these artifacts by age (children vs. adults) or sex. In my study, I have also considered information on the shape of skirts, on the way bark-cloth is used, and on the use of reeds, a material also used in skirt-making.

TABLE 7
Summary of an Operational Sequence: Two Men at Work, Followed by One Woman (1 July, 1979, Asiana, Papua New Guinea)

A. Stripping the Bark (men's work)

10:55 - 11:01	sharpen small sticks; strip first length of bark
11:02 - 11:04	fell tree (19 cm diameter)
11:04 - 11:06	measure first length of bark on fallen trunk
11:06 - 11:13	strip second length of bark
11:15 - 11:20	strip third length of bark
11:22 - 11:26	roll up the bark together
	cutting up/stripping three lengths of bark (average: 130 x 60 cm) total time: **31 minutes**

B. Peeling the Epidermis (men's work)

13:33 - 13:49	resting or slightly dynamic percussion using knife or machete
	total time: **16 minutes** (one bark)

C. Scraping Inside of Bark (men's work)

13:49 - 13:50	roll up the bark
13:50 - 14:16	resting percussion, toward self bark before pounding 129 x 59 cm
	total time: **27 minutes**

D. Pounding (women's work)

14:41 - 15:40 and 14:38 - 16:06	pound with a stick in strips of approximately 30 x 8 cm; 70 - 100 blows per minute time: 28 - 59 minutes **average: 44 minutes**
16:07 - 16:21	pounding with a stone on a small round wooden stick on *maro* folded double
16:21 - 16:32	on *maro* folded in four
16:32 - 16:46	on *maro* folded in eight
16:46 - 17:10	on *maro* folded in sixteen time: **63 minutes**

Total Time for a Cape

$\frac{31 \times 2}{3}$ + 27 + 44 + 63 = 155 minutes; rounded off to 2 hours 30 minutes
not counted: trips back and forth or drying

The material went from 129 x 59 cm to 164 x 118 cm, a 250% increase in area.

Source: Lemonnier 1984b.

Conclusions

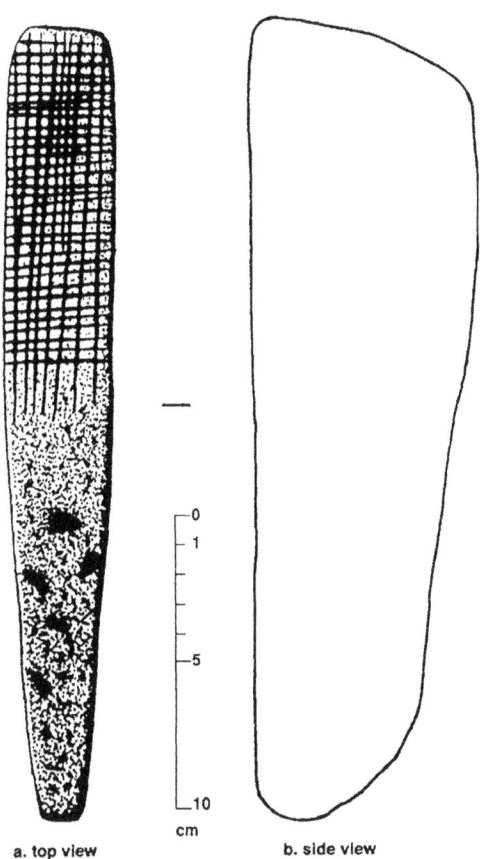

Figure 39. Anga stone beater for pounding bark (after Lemonnier 1984b).

Altogether, the use of pounded bark among the Anga can be summed up using 34 variables. The relationships among these variables has been systematically investigated. Subsequent reanalysis using factor analysis essentially replicated these initial results.

One result of the analysis is that the making and use of pounded bark by the Anga follows a basic feature of their social organization—the opposition between male and female worlds. This is particularly visible in the processing or use of the different plant species. Males are associated with the cultivated forms, whereas women are associated with the wild forms. The unidentified *Ficus*, "midzamanga," is a species in most cases used exclusively by females. When the *Ficus elastica* is not available in the environment, "midzamanga" is then cultivated, and females no longer use it, which is a way to "defeminize" it. Those few tribes in which males and females wear the same species as skirts and capes are also those lowland groups where male domination seems less strong, perhaps as a result of slightly different patterns of social and economic organization.

Not unexpectedly, the visible aspects of bark artifacts are correlated with ethnic identity. When other types of artifacts and body decorations are considered along with the bark items, the combined spatial distribution is clearly correlated with membership in particular ethnic (linguistic) groups. Whether this is done with intent is another question, although it is the case that for some of these artifacts, such as rattan and orchid belts, that voluntary marking of ethnic identity is commented upon by the informants. But let me sum up the results as follows: social representations of the making and use of pounded bark among the Anga are above all related to the general and fundamental opposition of male and female worlds.

However, if we now turn to other types of technological actions, the ethnic distribution of which has something to do with that of the bark industry, things become more complicated. For instance, the Anga commonly exchange bark-capes for vegetal salt (Godelier 1971; Lemonnier 1981). At the same time, the distribution of some features of the making or use of pounded-bark products more or less parallels some features of vegetal salt-making. Notably, the Baruya, producers of the "best" salt, do not produce their own bark-capes, nor their own bows, although suitable plant species are found in their environment and they possess the knowledge needed to make such items. Here, the explanation of these particular distributional patterns of specific technologies among the Anga lies in the necessities of intertribal trade: people are more prone to come and buy your particular product if they know that they can sell you some of their own, which you do not produce.

Now, if we put these facts together, and if we remember that some particular aspects (manufacture, use, shape) of pounded bark, bows, and vegetal salt follow the same pattern of ethnic distribution, we are either obligated to draw the unlikely conclusion that male/female relationships have something to do with the geographical pattern of intertribal trade, or to arrive at the improbable hypothesis that, at some level, as yet unknown relations of relations, totally out of reach for the time being, might influence both male/female relations and intertribal trade. I favor neither of these formulations, and assume instead that my description and analysis of Anga material culture is inadequate at present to understand its involvement in a system of meaning, that is, the general logic of technological choices among the

Anga. This may seem a disappointing result. I view it instead as a result that brings me back to more detailed field investigations. In this case, it also shows the limits of comparative studies, at least if they are done by only one ethnographer.

These comments notwithstanding, the Anga results have been obtained through simultaneous study of the physical as well as the more immediately informational aspects of several types of action on the material world, and not, as is often the case, by the study of isolated technologies, reduced to artifact decorations. It is this taking into account of both physical and informational functions of material culture, and of the systemic interrelationships among many technologies within one given set of societies, which led to these results. Finally, one should not be surprised to find that an anthropological study of material culture needs as much time as the study of any other social phenomenon. What is surprising is the implicit belief among some anthropologists that a limited set of features of a few isolated artifacts will reveal profound insights about the social relations involved in material culture: that the study of kinship, economy, or myths requires intensive fieldwork and reference to thousands of previous works, but that the study of technological systems, which are the social side of our everyday material life, can somehow be done just by "reading" the decorations on a few artifacts. As this is obviously an indefensible position, I assume that it is just another effect of the more or less unconscious ideological contempt in which many anthropologists hold any materialist approach to social phenomena.

Altogether, the Anga case proves that, although it is a fundamental aspect of the study of material culture, the study of the social representations of technologies is even less developed than that of other dimensions. This is a fact, and I see no way to avoid it, even though the void is a serious hindrance especially for archaeologists and for ethnologists interested in technological systems.

Spontaneous comments by informants on bark-making activities were very rare: for example, a Yoyue explained that a young boy could not wear midzamanga, whereas a young girl could; Baruya and Wantekia (a group of Baruya speakers) agree that young Baruya boys can wear either bark skirts or reed skirts, whereas young Wantekia boys should wear only reeds; a particular Baruya male head decoration can only be made from midzamanga bark; and the Baruya gave a very long statement, involving the possible pollution of male food by women, to explain why reeds destined to be used as female skirts cannot lie on bark capes. Since studying such statements in twelve different language groups is out of the question, in future fieldwork

I will focus my study of social representations of technologies on a single Anga group. I think that the study of technological discourse—a kind of "cognitive" approach—may be one way to improve our knowledge in these matters.

Whatever definition of cognitive anthropology is used, it has something to do with the way members of a social group perceive and organize in their minds the phenomena around them, be they social relations or the material world. One way to understand how social representations of an action on the material world are organized is to investigate the way people order these actions in and by their language. Once again, a prerequisite to such studies is that other aspects of technological systems be studied first. It is only when the materials, objects, and operational sequences of a particular technology are already known that this type of investigation is likely to be most successful.

The work of Lefébure (1978) illustrates one possible approach to such a study. Connotations are the various contextual meanings of a given linguistic sign, in addition to its basic meaning. Lefébure's assumption is that the connotations of the nouns people use when talking about a technology—in this case, the Berber vertical loom—are related to the way they perceive particular technological actions. In other words, lexical structures may be a clue to their perception of technological systems.

After describing the physical components (and use) of a vertical loom, Lefébure (1978) studied the naming system underlying the weaving terminology, as well as the etymology and connotations of weaving terms, and posited that

> the degree of specialization of a technological vocabulary and, complementarily, the more or less stressed coherence of its semantic structuring are a suitable guide to analyzing the relations between technologies and other social functions in a given group—for instance, in documenting the study of actors' specialization, that of the forms of transmission of knowledge, and that of the indigenous perception of work. [1978:118]

In the case of the Berber loom, the vocabulary is structured around pairs of logical oppositions that are all technologically pertinent (1978:121). For example, the term *iD* (warp) evokes verticality, whereas *aḍraf* (weft) evokes a furrow, or horizontality; similarly, *asaraw* and *aqrdaš* (other names for warp and weft, respectively) are opposed through reference to combed versus carded wool of which they are made.

Let us now consider the "motivation" (in Saussurian terms) underlying their weaving terminology, that is, the relations that link seemingly arbitrary linguistic signs to the technology. As shown by Lefébure, the motivation of these terms seems to correspond to the "tendency" (in Leroi-Gourhan's terms) that underlies their technological function. Thus, among the many possible terms one could apply to a warp, emphasizing for example its tension, verticality, strength, fineness, material, preparation, or continuity as a tightened thread, it is its verticality that has been emphasized in various Berber dialects, as well as in other languages (e.g., Greek, Latin, Arabic). And verticality in turn corresponds to a major "tendency" inherent in this particular technique of weaving, since an upright loom involves a better body position during work, the least amount of ground space, and the benefits of gravity when tightening the warp; all are technological advantages conferred by the verticality of the loom. The same kind of phenomenon is found in names for the spindle.

Only further studies of this type will tell us whether the correspondence between "motivation" and "tendency" in Berber weaving technology also holds in other technologies and in other societies. Whatever the case, if the linguistic denotation of technological actions has something to do with their perception, such studies may provide an important indirect access to the various mental representations and operations associated with technological actions.

Other possible "cognitive" approaches might deal with the mental procedures of decision-making involved in technological actions. By looking at people at work and asking them questions, Young, for example, has recently provided accounts of experimental stone-tool making (Young and Bonnichsen 1984), as well as skinworking and basket-making techniques (Young 1985). Besides describing and analyzing operational sequences, these studies have focused mainly on psychological aspects of cognition and behavior; namely, the existence of a craftperson's

> grammar (or code), composed of relatively stable repertoires and rules, to create a great variety of different artifacts depending on the circumstances and materials available. [Young and Bonnichsen 1984:152]

Although the sociological aspects of these cognition and decision-making investigations are unfortunately missing in these otherwise very precise and useful studies, they point to mental procedures the sequencing and consciousness of which may vary from one culture to another. Comparative studies are missing here, as in other matters, which would allow us to make general statements on questions such as repetitiveness vs. singularity in operational sequences.

As I have already mentioned, the indirect study of the technological principles involved in given actions might be another way to help us circumscribe the domain of technological choices. In effect, besides the use of materials and tools, which obviously varies from one society to another, and the performance of particular movements, which unfortunately are still beyond our scope, the recourse to particular physical principles (i.e., of particular technological knowledge) may, or may not, be subject to cultural variation. In this case, the use or non-use of such technological principles would be another systemic feature of material culture. Up to now, this is still a totally new field of research (but see Pelegrin 1988 for a tentative and useful exploration of the mental processes involved in a technological task).

In the end, this book is not only programmatic; it may even seem unfinished. The reason is that this particular field of anthropology is still in its genesis, and it will continue searching for appropriate methods and defining its proper domain of research for many years to come. But at least the central issue has been identified: technological systems are social productions, and for this reason they are appropriate subjects for anthropological investigation; no facet of human action on the material world should be ignored. The physical and informational dimensions of technology are embedded in each other, simply because societies are always manipulating symbols whether in designing airplanes or in rocking babies.

The critical data base may still be out of reach, mainly because appropriate field studies have not been conducted, but the goal of an ethnology of technology is nevertheless worth the effort. Ultimately, at issue here is the reintroduction of the material aspects of human life back into the domain of anthropology, a necessity which unfortunately remains less than obvious to many anthropologists. We must renew our focus on the relation between the forces of production and the social relations of production—in other words, between "technology," "society," and "culture"—as a process that is mediated by particular social representations or "technological knowledge."

Such a renewed focus would permit us to relate the ethnology of technology to classical, and much more developed, themes in anthropology. This can be achieved not only by encouraging cooperation among archaeologists, historians, and ethnologists, but also by drawing together into the study of technology the insights of such disciplines as cognitive anthropology, psychology, and ethnolinguistics.

Together with more fieldwork, the cooperative efforts of specialists in these various disciplines provide the surest avenue toward a better understanding of the exciting issues of whether and how technological systems also constitute systems of meaning, and the degree to which social choices influence the transformation of technological systems and societies.

Notes

1. I am indebted to Jean-François Quilici, a civil engineer, for this particular systemic approach to the classification of technological phenomena.

2. Unfortunately, Driver and Massey (1957:294–310) provide only nine maps on housing, which is a lot by present standards in ethnology for issues concerning material culture, but which is still inadequate as a basis for an anthropology of technological systems. This is not the place to discuss the heterogeneity of the "traits" they selected for mapping, nor the way they were used.

3. The non-use of sea water by people living in the middle of the island is obviously not the result of a choice, but simply its unavailability.

4. These paragraphs are a revised and shortened version of a paper entitled "Bark Capes, Arrowheads and Concorde: On Social Representations of Technology" (Lemonnier 1989).

5. The image of the designer as joyous improvisor or flying mad man is an historical untruth. The Wright brothers, for example, made detailed studies of Octave Chanute's work (designer of many gliders), and they had many years of experience with biplane kites (1899) and gliders (1900–1902) before they succeeded in their first flight "with engine, of long distance, and controlled, taking off from a flat terrain" in December, 1903 (Hardt 1980:34–39). It is on the drawing-boards and in the patents that aviation was born (Culick 1979; Chadeau 1986).

6. Even the Northrop B2 bomber (first public appearance, November, 1988) might have its series production cancelled because of its billion dollar price tag. It is noteworthy that it is the stealth materials and construction which are responsible for its high cost, not the "flying wing" concept.

7. Recently, Delaporte (1988) developed such an approach in the study of Lapp costume.

8. Although the SCI's research focus is on technological systems in industrial settings, or on the transfer of modern technologies to non- or semi-industrial societies, its members' aims and methods are distinctly those of an anthropology of technology as defined here (see, for example, Akrich 1987; Callon 1986; Callon and Latour 1985; Latour and Woolgar 1986).

9. My comments on the Cambridge School have no other motive than to encourage these scholars to expand, not narrow, their theoretical perspective. Technology, like other social productions, cannot simply be reduced to just a single dimension. All hands are needed to build a body of anthropological theory for material culture that encompasses all of its dimensions, whether they relate to physical actions on matter, systems of meaning, or social change. Now is not the time for hasty generalizations, but rather for minute studies of particular problems in the widest theoretical context possible.

10. Since these pages were first written (1986), Descola's (1987) *La Nature Domestique*—a description and analysis of Jivaro Achuar symbolic views of hunting and gardening techniques—has been published. Although this book covers many different issues, it is the first study ever published, to my knowledge, on the anthropology of technology as conceived of here.

References Cited

Abel, F.
1984 *Ejen 'Pétrir': Produire de la Nourriture—et du Sens? Techniques et Culture* (n.s.) 3:65–80.

Akrich, M.
1987 Comment Décrire les Objets Techniques. *Techniques et Culture* (n.s.) 9:81–119.

Angelucci, E., and P. Matricardi (Eds.)
1983 *L'Encyclopédie des Avions Civils du Monde des Origines à nos Jours.* Fernand Nathan, Paris.

Armacost, M. H.
1985 The Thor-Jupiter Controversy. In *The Social Shaping of Technology*, edited by D. MacKenzie and J. Wajcman, pp. 252–62. Open University Press, Milton Keynes, England.

Aymard, A.
1959 Stagnation Technique et Esclavage. In *Histoire Général du Travail* edited by L.-H. Parias, pp. 371–77. Nouvelle Librairie de France, Paris.

Bahuchet, S.
1985 *Les Pygmées Aka et la Forêt Centrafricaine: Ethnologie Écologique.* Société d'Etudes Linguistiques et Anthropologiques de France, Paris.

Bazin, M., and C. Bromberger
1982 *Gîlan et Azarbâyjjan Oriental: Cartes et Documents Ethnographiques.* Editions Recherches sur les Civilisations, A.D.P.F., Paris.

Beneveniste, E.
1974 *Problèmes de Linguistique Générale*, Vol. 2. Gallimard, Paris.

Biasutti, R.
  1952 *La Casa Rurale della Toscana.* Centro di Studi per la Geografia Etnologia (Ricerche sulla Dimore Rurali in Italia [CNR]), Firenze.

Bloch, M.
  1935 Avènement et Conquête du Moulin à Eau. *Annales d'Histoire Economique et Sociale* 7:538–63.

Bogatyrev, P.
  1971 *The Functions of Folk Costume in Moravian Slovakia.* Mouton, The Hague.

Braudel, F.
  1977 *Afterthoughts on Material Civilization and Capitalism.* Johns Hopkins University Press, Baltimore.

Bresson, F.
  1987 Les Fonctions de Représentation et de Communication. In *Psychologie*, edited by J. Piaget, P. Mounoud, and J.-P. Bronckart, pp. 932–82. Gallimard, Paris.

Bril, B.
  1986 The Acquisition of an Everyday Technical Motor Skill: The Pounding of Cereals in Mali (Africa). In *Themes in Motor Development*, edited by M. G. Wade and H. T. A. Whiting, pp. 315–26. Martinus Nijhoff, Dordrecht.

Bril, B., and C. Sabatier
  1986 The Cultural Context of Motor Development: Postural Manipulations in the Daily Life of Bambara Babies (Mali). *International Journal of Behavioral Development* 9:439–53.

Bromberger, C.
  1979 Technologie et Analyse Sémantique des Objets. *L'Homme* 19(1):105–40.

Bromberger, C., D. Dosseto, and T.-K. Schippers
  1982–83 L'Ethnocartographie en Europe: Coups d'Oeil Rétrospectifs et Questions Ouvertes. *Technologies, Idéologies, Pratiques* 4(1–4):15–39 (Numéro Spécial, "L'Ethnocartographie en Europe").

Brown, G.
  1910 *Melanesians and Polynesians: Their Life-Histories Described and Compared.* Macmillan and Company, London.

Calame-Griaule, G.
  1977 Pour l'Étude des Gestes Narratifs. In *Langage et Culture Africaine*, edited by G. Calame-Griaule, pp. 303–58. Gallimard, Paris.

Callon, M.
　1986　Eléments pour une Sociologie de la Traduction: La Domestication des Coquilles St. Jacques et des Marins Pêcheurs dans la Baie de St. Brieuc. Ecole Nationale Supérieure des Mines de Paris, Paris.

Callon, M., and B. Latour
　1985　Les Scientifiques et leurs Alliés. Pandore, Paris.

Chadeau, E.
　1986　Poids des Filières Socio-Culturelles et Nature de l'Invention: L'Aéroplane en France jusqu'en 1908. L'Année Sociologique 36:93–112.

Chamoux, M.-N.
　1983　La Division des Savoir-Faire Textiles Entre Indiens et Métis dans la Sierra de Puebla (Mexique). Techniques et Culture (n.s.) 2:99–124.

Chapanis, A.
　1965　Research Technique in Human Engineering. Johns Hopkins University Press, Baltimore.

Coles, J.
　1973　Archaeology by Experiment. Hutchinson University Library, London.

Conklin, H.
　1982　Ethnoarchaeology: An Ethnographer's Viewpoint. In Ethnography by Archaeologists: 1978 Proceedings of the American Ethnological Society, pp. 11–17. American Ethnological Society, Washington, DC.

Cresswell, R.
　1968　Le Geste Manuel Associé au Langage. Langages 10:119–27.
　1976　Avant-Propos. Techniques et Culture 1:5–6.
　1983　Transferts de Techniques et Chaînes Opératoires. Techniques et Cultures (n.s.) 2:143–63.

Culick, F. E. C.
　1979　The Origins of the First Powered, Man-Carrying Airplane. Scientific American 241(1):86–100.

Delaporte, Y.
　1988　Les Costumes du Sud de la Laponie: Organisation et Désorganisation d'un Système Symbolique. Techniques et Culture (n.s.) 12:1–19.

Demesse, L.
　1978　Changements Techno-Economiques et Sociaux Chez les Pygmées Babinga (Nord Congo et Sud Centrafrique). Société d'Etudes Linguistiques et Anthropologiques de France, Paris (2 volumes).
　1980　Techniques et Économie des Pygmées Babinga. Institut d'Ethnologie, Paris.

Descola, P.
　1987　*La Nature Domestique. Symbolisme et Praxis dans l'Ecologie des Achuars.* Editions de la Maison des Sciences de l'Homme, Paris.

Devereux, G.
　1956　Normal and Abnormal: The Key Problem of Psychiatric Anthropology. In *Some Uses of Anthropology: Theoretical and Applied*, edited by J. B. Casagrande and T. Gladwin, pp. 23–48. Anthropological Society of Washington, Washington, D.C.

Digard, J.-P.
　1979　La Technologie en Anthropologie: Fin de Parcours ou Nouveau Souffle? *L'Homme* 19(1):73–104.

Driver, H. E., and W. C. Massey
　1957　Comparative Studies of North American Indians. *Transactions of the American Philosophical Society* 67:165–456.

Durkheim, E.
　1950　*Rules of Sociological Method* (8th edition). Free Press, New York.

Efron, D.
　1941　*Gesture and Environment: A Tentative Study of Some of the Spatio-Temporal and "Linguistic" Aspects of the Gestural Behaviour of Eastern Jews and Southern Italians in New York City, Living Under Similar as Well as Different Environmental Conditions.* King's Crown Press, New York (reprinted 1972, Mouton, The Hague).

Eibl-Eibesfeldt, I.
　1972　*Grundriss der Vergleichenden Verhaltensforschung.* Piper Verlag, Munchen.

Esparragoza, M.-E.
　1983　Opérations et Phases d'une Activité Matérielle: L'Étude d'une Fabrication Culinaire. *Techniques et Culture* (n.s.) 1:171–86.

Fallows, J.
　1985　The American Army and the M-16 Rifle. In *The Social Shaping of Technology*, edited by D. MacKenzie and J. Wajcman, pp. 238–51. Open University Press, Milton Keynes, England.

Febvre, L.
　1935　Réflexion sur l'Histoire des Techniques. *Annales d'Histoire Économique et Sociale* 7:531–35.

Finley, M. I.
　1965　Technical Innovation and Economic Progress in the Ancient World. *The Economic History Review* (2nd series) 18:29–45.

Geistdoerfer, A.
　1983　Fonctions Spécifiques de l'Pêche dans une Production Halieutique. *Techniques et Culture* (n.s.) 2:87–98.

Gille, B.
　1966　*Engineers of the Renaissance*. M.I.T. Press, Cambridge, Mass.
　1970　L'Évolution de la Technique Sidérurgique. Esquisse d'un Schéma. *Revue d'Histoire des Mines et de la Métallurgie* 2:121–226.
　1980　*Les Mécaniciens Grecs: La Naissance de la Technologie*. Le Seuil, Paris.

Gille, B. (Ed.)
　1978　*Histoire des Techniques: Technique et Civilisation, Technique et Sciences*. Gallimard, Paris.

Godelier, M.
　1971　Salt Currency and the Circulation of Commodities Among the Baruya of New Guinea. In *Studies in Economic Anthropology*, edited by G. Dalton, pp. 52–73. American Anthropological Association, Anthropological Studies 7.
　1973　Outils de Pierre, Outils d'Acier Chez les Baruya de Nouvelle-Guinée. *L'Homme* 13(3):187–220.
　1977　*Perspectives in Marxist Anthropology*. Cambridge University Press, Cambridge.
　1986　*The Mental and the Material: Thought, Economy and Society*. Verso, London.

Hardt, C.
　1980　Cerfs-Volants et Planeurs. In *Encyclopédie de l'Aviation*, edited by D. Mondey, pp. 22–39. Compagnie Internationale du Livre, Paris.

Haudricourt, A.-G.
　1962　Domestication des Animaux, Culture des Plantes et Traitement d'Autrui. *L'Homme* 2(1):40–50.
　1968　La Technologie Culturelle: Essai de Méthodologie. In *Ethnologie Générale*, edited by J. Poirier, pp. 731–822. Gallimard, Paris.

Heider, K.
　1970　*The Dugum Dani: A Papuan Culture in the Highlands of West New Guinea*. Viking Fund Publication in Anthropology 49, Wenner-Gren Foundation for Anthropological Research, New York.

Hodder, I.
　1982　*Symbols in Action: Ethnoarchaeological Studies of Material Culture*. Cambridge University Press, Cambridge.

Hooven, F. J.
　1978　The Wright Brothers' Flight-Control System. *Scientific American* 239(5):167–84.

Koechlin, B.
   1972   A Propos de Trois Systèmes de Notation des Positions et Mouvements du Corps Humain Susceptibles d'Intéresser l'Ethnologue. In *Langues et Techniques, Nature et Société, Tome 2, Approche Ethnologique, Approche Naturaliste*, edited by J. Thomas and L. Bernot, pp. 157–84. Klincksiek, Paris.

Koechlin, B., and J. Matras
   1971   Pour une Ethno-Technologie: Eléments d'un Manuel de Technologie Culturelle (Un Prototype de Grille Documentaire). *Asie du Sud-Est et Monde Insulindien* (Bulletin du Centre de Documentation et de Recherche sur l'Asie du Sud-Est et le Monde Insulindien) 2–3:7–172.

Kroeber, A. L.
   1957   *Style and Civilization.* Cornell University Press, Ithaca, New York.

Latour, B., and S. Woolgar
   1986   *Laboratory Life. The Social Construction of Scientific Facts.* Princeton University Press, Princeton.

Lee, R.
   1969   !Kung Bushman Subsistence: An Input-Output Analysis. In *Environment and Cultural Behavior*, edited by A. P. Vayda, pp. 47–79. Natural History Press, New York.
   1979   *The !Kung San: Men, Women, and Work in a Foraging Society.* Cambridge University Press, Cambridge.

Lefébure, C.
   1978   Linguistique et Technologie Culturelle: L'Exemple du Métier à Tisser Vertical Berbère. *Techniques et Culture* 3:84–148.

Lefebvre des Noëttes, Cdt.
   1931   *L'Attelage. Le Cheval de Selle à travers les Âges. Contribution à l'Histoire de l'Esclavage.* Picard, Paris.

Lemonnier, P.
   1976   La Description des Chaînes Opératoires: Contribution à l'Analyse des Systèmes Techniques. *Techniques et Culture* 1:100–151.
   1980   *Les Salines de l'Ouest. Logique Technique, Logique Sociale.* Editions de la Maison des Sciences de l'Homme, Paris.
   1981   Le Commerce Inter-Tribal des Anga de Nouvelle-Guinée. *Journal de la Société des Océanistes* 37(70–71):39–75.
   1982   Jardins Anga (Nouvelle-Guinée). *Journal d'Agriculture Traditionnelle et de Botanique Appliquée* 29(3–4):227–45.
   1983   La Description des Systèmes Techniques: Une Urgence en Technologie Culturelle. *Techniques et Culture* (n.s.) 1:11–26.
   1984a   La Production de Sel Végétal Chez les Anga (Papouasie Nouvelle-Guinée). *Journal d'Agriculture Traditionnelle et de Botanique Appliquée* 31(1–2):71–126.
   1984b   L'Écorce Battue Chez les Anga de Nouvelle-Guinée. *Techniques et Culture* (n.s.) 4:127–75.

1986 The Study of Material Culture Today: Toward an Anthropology of Technical Systems. *Journal of Anthropological Archaeology* 5:147–86.
1987 Le Sens des Flèches: Culture Matérielle et Identité Ethnique Chez les Anga de Nouvelle-Guinée. In *De la Voute Céleste au Terrain, du Jardin au Foyer. Mosaïque Sociographique*, edited by B. Koechlin, F. Sigaut, J. M. C. Thomas, and G. Toffin, pp. 573–95. Editions de l'Ecole des Hautes Etudes en Sciences Sociales, Paris.
1989 Bark Capes, Arrowheads and Concorde: On Social Representations of Technology. In *The Meaning of Things. Material Culture and Symbolic Expression*, edited by I. Hodder, pp. 156–71. Unwin Hyman, London.

Leroi-Gourhan, A.
1943 *Evolution et Techniques: L'Homme et la Matière*. Albin Michel, Paris (reprinted 1971).
1945 *Evolution et Techniques: Milieu et Techniques*. Albin Michel, Paris (reprinted 1973).

Lévi-Strauss, C.
1976 *Structural Anthropology*, Volume 2. University of Chicago Press, Chicago.
1985 *The View from Afar*. Basic Books, New York.

Mac McClellan, J.
1985 The Odd Squad: Mitsubishi MU-2, an Odd-Looking Machine. *Flying* 112(4):40–42.

Malinowski, B.
1935 *Coral Gardens and their Magic: A Study of the Methods of Tilling the Soil and of Agricultural Rites in the Trobriand Islands*. George Allen and Unwin, London.

Marx, K.
1976 *Capital: A Critique of Political Economy*, Vol. 1. Penguin Books, Harmondsworth, England.

Mauss, M.
1935 Les Techniques du Corps. *Journal de Psychologie* 32(3–4):271–93 (English translation in Mauss, M., 1979, *Sociology and Psychology*. Routledge and Kegan Paul, London).
1947 *Manuel d'Ethnographie*. Payot, Paris.

Meillassoux, C.
1964 *Anthropologie Économique des Gouro de Côte d'Ivoire: De l'Économie de Subsistance à l'Agriculture Commerciale*. Mouton, The Hague.
1967 Recherche d'un Niveau de Détermination dans la Société Cynégétique. *L'Homme et la Société* 6:95–106.
1981 *Maiden, Meal and Money: Capitalism and the Domestic Community*. Cambridge University Press, New York.

Mumford, L..
1934 *Technics and Civilization*. G. Routledge and Sons, London.

Needham, J.
1969 *The Grand Titration: Science and Society in East and West*. Allen and Unwin, London.
1970 *Clerks and Craftsmen in China and the West: Lectures and Addresses on the History of Science and Technology*. Cambridge University Press, Cambridge.

Ogibenin, B. L.
1971 Petr Bogatyrev and Structural Ethnography. In *The Functions of Folk Costume in Moravian Slovakia*, edited by P. Bogatyrev, pp. 8–32. Mouton, The Hague.

Owen, K.
1982 *Concorde: New Shape in the Sky*. Jane's Publishing Company, London.

Pelegrin, J.
1988 A Framework for Analyzing Prehistoric Stone Tool Manufacture and a Tentative Application to some Early Lithic Industries. Unpublished manuscript presented in a colloquium "L'Outil Chez l'Homme et l'Animal," organized by Foundation FYSSEN, Versailles, Nov. 26–28, 1988.

Pelosse, J.-L.
1956 Contribution à l'Étude des Usages Corporels Traditionnels: Eléments d'Analyse des Mouvements Moteurs Socialisés en Ethnologie, Mouvements Segmentaires. *Revue Internationale d'Ethno-Psychologie Normale et Pathologique* 2:3–31.
1981 Analyse Gestuelle de Trois Procédés de Tricotage Pratiqués en Europe. *Geste et Image* 2:13–43.

Polhemus, T. (Ed.)
1978 *Social Aspects of the Human Body*. Penguin Books, Harmondsworth, England.

Propp, V. J.
1968 *Morphology of the Folktale*. University of Texas Press, Austin.

Quilici-Pacaud, J.-F.
1977 Réflexions Comparatives sur la Dynamique et l'Architecture des Véhicules (Automobiles, Avions, Bateaux). *Ingénieurs de l'Automobile* (Revue de la Société des Ingénieurs de l'Automobile), pp. 519–24 (August).
1987 Hommage à André Leroi-Gourhan. Lettre d'un Technicien à ses Amis Ethnologues. *Techniques et Culture* (n.s.) 10:45–59.
1989 Technologie et Systémique. In *Perspectives Systémiques*, edited by B. Paulré, pp. 148–61. L'Interdisciplinaire, Limonest.

Rappaport, R.
   1968   *Pigs for the Ancestors: Ritual in the Ecology of a New Guinea People.* Yale University Press, New Haven.

Reynolds, B.
   1983   The Relevance of Material Culture to Anthropology. *Journal of the Anthropological Society of Oxford* 14(2):209–217.

Reynolds, P.
   1978   *Iron Age Farm: The Butser Experiment.* British Museum Publications, London.

Rosenberg, N.
   1963   Technological Change in the Machine-Tool Industry, 1840–1910. *Journal of Economic History* 23(4):414–43.

Rostow, W. W.
   1975   *How it all Began: Origins of the Modern Economy.* McGraw-Hill, New York.

Sackett, J. R.
   1982   Approaches to Style in Lithic Archaeology. *Journal of Anthropological Archaeology* 1:59–112.

Salmona, M.
   1983   Transformations Technologiques et Vulgarisation Scientifique: Histoire des Apprentissages Précoces, Imaginaire et Activité Technique. *Techniques et Culture* (n.s.) 1:71–99.

Steensberg, A.
   1980   *New Guinea Gardens: A Study of Husbandry with Parallels in Prehistoric Europe.* Academic Press, New York.

Schuhl, P.-M.
   1969   *Machinisme et Philosophie.* Presses Universitaires de France, Paris.

Sigaut, F.
   1980   La Technologie: Obstacles et Promesses. *Journées d'Études de la Société d'Ethnologie Française,* Marseille (communication, December 1980).
   1984   Essai d'Identification des Instruments à Bras de Travail du Sol. *Cahier ORSTOM* (série Sciences Humaines) 20(3–4):359–74.
   1985   *Introduction à l'Évolution Technique des Agricultures Européennes Avant l'Époque Industrielle.* École des Hautes Études en Sciences Sociales (Centre de Recherche Historique), Paris.

Sorenson, E. R.
   1972   Socio-Ecological Change Among the Fore of New Guinea. *Current Anthropology* 13:349–83.

Swanson, E.
  1975 *Lithic Technology: Making and Using Stone Tools.* Mouton, The Hague.

Sweetman, B.
  1985 Avions Indetectables. *Interavia* 11:1217–22.

Taylor, M. J. H.
  1981 *Fantastic Flying Machines.* Jane's Publishing Company, London.

Techniques et Culture (E. R. 191 du Centre National de la Recherche Scientifique, Paris.)
  1977 Unpublished Internal Seminar.

Terray, E.
  1972 *Marxism and "Primitive" Societies: Two Studies.* Monthly Review Press, New York.

Thurber, M.
  1985 The Odd Squad: C137. *Flying* 112(4):37–40.

Tilley, C.
  1986 Interpreting Material Culture. In *Proceedings of the World Archaeological Congress, II. Material Culture and Symbolic Expression*, edited by I. Hodder, pp. 1–21. University of Southampton, Southampton.

Turnbull, C.
  1966 *Wayward Servants.* Eyre, Spottishwoode, London.

Watson, J. B.
  1965 The Significance of a Recent Ecological Change in the Central Highlands of New Guinea. *Journal of the Polynesian Society* 74:438–50.
  1977 Pigs, Fodder, and the Jones Effect in Postipomoean New Guinea. *Ethnology* 16(1):57–70.

White, Leslie
  1959 *The Evolution of Culture: The Development of Civilization to the Fall of Rome.* McGraw-Hill, New York.

White, Lynn, Jr.
  1962 *Medieval Technology and Social Change.* Oxford University Press, Oxford.

Wiessner, P.
  1982 Risk, Reciprocity and Social Influence on !Kung San Economics. In *Politics and History in Band Societies*, edited by E. Leacock and R. B. Lee, pp. 61–84. Cambridge University Press, Cambridge.
  1983 Style and Social Information in Kalahari San Projectile Points. *American Antiquity* 48(2):253–76.
  1984 Reconsidering the Behavioral Basis for Style: A Case Study Among the Kalahari San. *Journal of Anthropological Archaeology* 3:190–234.

1989 Style and Changing Relations Between the Individual and Society. In *The Meaning of Things. Material Culture and Symbolic Expression*, edited by I. Hodder, pp. 56–63. Unwin Hyman, London.

Wobst, H. M.
1977 Stylistic Behavior and Information Exchange. In *Papers for the Director: Research Essays in Honor of James B. Griffin*, edited by C. E. Cleland, pp. 317–42. University of Michigan Museum of Anthropology, Anthropological Paper 61, Ann Arbor, Michigan.

Wooldridge, D. E.
1983 *Winged Wonders: The Story of the Flying Wing*. National Air and Space Museum, Smithsonian Institution Press, Washington, DC.

Yengoyan, A.
1986 Theory in Anthropology: On the Demise of the Concept of Culture. *Comparative Studies in Society and History* 28(2):368–74.

Young, D. E.
1985 The Need for a Cognitive Approach to the Study of Material Culture. *Culture* 5(2):53–67.

Young, D. E., and R. Bonnichsen
1984 *Understanding Stone Tools: A Cognitive Approach*. Peopling of the Americas, Process Series, Vol. 1. Center for the Study of Early Man, University of Maine at Orono.

CPSIA information can be obtained
at www.ICGtesting.com
Printed in the USA
FSHW011444130122
87437FS

9 780915 703302